World's Most Dangerous

a history of the Columbia River Bar, its pilots and their equipment

by Michael E. Haglund

Columbia River Maritime Museum
Astoria, Oregon 2011

Published by
Columbia River Maritime Museum
1792 Marine Drive
Astoria, Oregon 97103

Authored by Michael E. Haglund

Illustrated by Eric Baker

Designed by Whole Brain Creative, Inc.
Tiffany Estes and Amy Bornstein

Printed by Printgraphics

To the courage and dedication
of the Columbia River Bar Pilots,
past, present and future.

World's Most
Dangerous

ACKNOWLEDGEMENTS

THE WRITING OF ANY BOOK requires the help and support of many. I am particularly indebted to the Columbia River Bar Pilots for hiring me in 1998 as their legal counsel. It was in preparing for a major rate proceeding to secure funding for the helicopter/fast boat combination transferring pilots today that I was first exposed to the extraordinary history of the bar pilots and began to ask the question: Is the Columbia River Bar really the most dangerous in the world? Captains Gary Lewin, Roger Nelson, and Thron Riggs were particularly helpful in answering hundreds of questions and continually pointing me in the right direction for additional research.

It was Thron Riggs who suggested that the Columbia River Maritime Museum might be the ideal publisher and that has indeed been the case. Executive director Sam Johnson helped me develop the organizational structure of the book and deputy director Dave Pearson secured both an outstanding illustrator in Eric Baker and a gifted designer in Tiffany Estes. Dave also assisted in finding and selecting the photographs for this book. Museum store manager Blue Anderson deserves special mention for her gentle, but consistent prodding to keep this project moving toward the printer.

Illustrations and charts can do much to cut through the fog of complicated subject matter. This book has benefited greatly from the talent of illustrator Eric Baker, who brought his years of experience with Oregon's largest newspaper to the brilliant watercolor illustrations throughout. His renderings of multiple charts bring a consistency to this book's cartography that should be of help to the reader.

My work on this book occurred in widely separated spurts over a period of five years. When writer's block ensued, my wife Melissa was there with an encouraging word and a proofreader's eye. Our daughter Molly spent most of one summer performing research at multiple maritime museums and libraries, the Clatsop County Historical Society, and the Oregon Historical Society. And our daughter Christina helped out with some of the fact checking and photographic research.

Boat operator Don Nelson, a veteran of more bar crossings than any single human in history, was a great resource in nailing down the facts of multiple events over the last 60 years. Bar pilot Captain Robert Johnson provided helpful edits and corrections and museum volunteer Margaret Hines, a professional editor, brought an expertise that was much appreciated.

I must also acknowledge the late Russell Dark, a journalist who retired to Astoria in 1968 and then devoted great effort to researching the history of the Columbia River Bar Pilots. He died in 1980 at age 76 with his work unfinished, but he left an unpublished manuscript entitled *Graveyard Passage* that was a rich factual resource.

And special thanks to Professor Tuba Ozkan-Haller of Oregon State University's College of Oceanic and Atmospheric Sciences, who explained the remarkable physics of the Columbia River Bar and whose calculations definitively answer the central question of this book.

Astoria is truly blessed to have one of America's finest maritime museums in a beautiful riverfront setting on the Columbia River. I am honored to dedicate the proceeds of this book to the Columbia River Maritime Museum in the hopes of contributing to its mission to preserve the maritime history of the Great River of the West.

TABLE OF CONTENTS

INTRODUCTION

THIS IS A BOOK OF THREE STORIES. Accordingly, it is organized into three parts.

The first part examines the unique geologic history of the Columbia River Bar and the three different eras of its navigational history: pre-jetty, jetty construction, and post-jetty.

This first section systematically examines and confirms that the Columbia River Bar is the most dangerous entrance to a commercial waterway in the world. In the pre-jetty era, the evidence of the bar's predominant reputation for danger is found in the observations of early explorers who described the unrivaled treachery of the entrance to the Columbia River, the British Admiralty's sailing directions published during this period, and the remarkable number of ships wrecked on the bar. Nowhere on the planet were the perils to the approaching mariner so starkly stated and the need for a pilot accorded so much emphasis.

Some have suggested that the construction of jetties at the Columbia River entrance and regular dredging have tamed the bar and called into question the observations predating the completion of the north jetty in 1917. But this old saw is laid to rest in two chapters.

One chapter documents the remarkable history of jetty construction spanning 32 years that produced jetties of unparalleled size when first constructed, and remains a work in progress. Contrary to the claims of various promoters of commerce on the Columbia River, the U.S. Army Corps of Engineers has never declared the fearsome bar to be gone. Rather, the Corps has produced a channel through the bar that requires constant maintenance, has evolved significantly over the last century, and depends upon the unique skill set of the Columbia River Bar Pilots to be functional in rough conditions. Indeed, one of the most interesting facets of the bar's history is the symbiotic relationship between the Corps' continuing efforts to engineer and maintain a channel through the Columbia River Bar and the navigational prowess of the Columbia River Bar Pilots.

Another chapter compares the navigational challenge of the modern Columbia River Bar to the other top candidates from every continent. The evidence is scientific, not anecdotal, and examines the bathymetry, currents, and waves encountered in each of these coastal inlets.

Suffice it to say here that the Columbia River is in the top rank of world waterway entrances in terms of incoming ocean swell, river ebb current, and winter wave heights. While not the worst in every category, what distinguishes the Columbia River Bar as the world's most dangerous is its perilous combination of wind and wave meeting mighty river at a point where the approaching or departing ship must execute a turn between two wreck-strewn spits on a lengthy bar.

The bottom line: nowhere in the world is a ship more likely to meet taller waves at a waterway entrance.

This book's second part is the story of the Columbia River Bar Pilots, who were so important to commerce in the region that they were among the first professionals of any kind to be licensed by Oregon's territorial government more than a decade before statehood in

1859. Central to this story is the requirement that bar pilots be experienced oceangoing shipmasters. What began as a process of natural selection – only experienced captains found success on the bar – ultimately became a policy adopted by that most famous of bar pilots, Captain George Flavel, and has been embedded in regulations adopted by Oregon's Pilotage Commission since at least 1940.

This part begins with a chapter devoted to the history of the Columbia River Bar Pilots as an organization and then proceeds to a collection of sea story chapters about the bar and its pilots. Because it is often easier to understand the past through the lens of the present, these chapters unfold in reverse chronological order. The first involves an account from 2002 involving the modern bulk carrier *Tai Shan Hai* that almost wrecked on Peacock Spit. Were it not for the expertise of three bar pilots and the capability of the only pilotage-dedicated helicopter in North America, this near miss would have become a major catastrophe. The *Tai Shan Hai* would have broken up in the pounding winter surf and spilled her huge stores of fuel, generating headlines around the world and confirming the madness of approaching the Columbia River Bar without a pilot.

The sea story chapters go back over more than 150 years to Captain George Flavel's famous but unsuccessful efforts to prevent the loss of the steam-powered sailing vessel *General Warren* in 1852. This section concludes with a chapter chronicling the 17 bar pilots and deckhands who have lost their lives in the line of duty.

The book's third part covers the unique story of bar pilot transfer technology, the equipment that has evolved over 200 years to carry bar pilots to and from their assigned vessels in the open ocean. While Chinook Chief Concomly led ships in by canoe from a haling point inside the bar, the first bar pilots to cross the bar aboard the ship being piloted met those vessels aboard a sailing schooner rigged with a small rowboat to make the pilot transfer.

For well over a century, a rowboat or pulling boat launched from a larger vessel was used to transfer bar pilots to and from ships in the open sea off the mouth of the Columbia River. During this period, the pilot vessels carrying the transfer boat changed from sailing schooner to steam tugboat, but the use of a wooden pulling boat powered by two skilled and hardy oarsmen remained the mainstay in boarding and disembarking bar pilots offshore.

The first quantum leap change in pilot transfer technology came in the form of a specially designed and modified German rescue boat, which was delivered to Astoria in 1967, the incomparable *Peacock*. Diesel-powered, steel-hulled and self-righting, she was the only pilot boat of her kind in the world. Instead of a rowboat, the *Peacock* carried a motorized aluminum daughter boat secured in a cradle and then launched down a ways off her stern for pilot transfers in the roughest of conditions. With a boat operator at the controls, the bar pilot waited for that moment when the daughter boat was at the crest of a swell adjacent to the pilot ladder before making the leap to the ladder and scrambling topside to begin the pilotage assignment.

The *Peacock* era spanned more than three decades, well beyond her projected useful life. What had been a state-of-the-art pilot transfer system during the *Peacock's* heyday had become dangerously obsolete with the rapid increase in the size of the world's shipping fleet in the 1980s and 1990s. The bar pilots knew the *Peacock* needed replacement, but with what? The organization spent nearly 15 years studying how best to make the transition to an entirely new generation of pilot transport including a helicopter and a fast, maneuverable cutter-type pilot boat. True to their more than 150-year history of repeatedly embracing new pilot transportation technologies that both improved pilot, ship, and environmental safety and enhanced commerce on the Columbia River, the bar pilots in 1999–2000 successfully introduced the first dedicated helicopter operation in North America combined with what is now a pair of fast, aluminum jet-powered pilot boats. At present, two-thirds of the pilot transfers offshore are accomplished by helicopter and the other third by pilot boat.

Although divided into three parts, each of this book's three stories is necessarily intertwined. The science underlying the proof of the bar's most treacherous status is best illustrated through the stories of ships lost and saved throughout history, both with and without bar pilots. And a critical examination of the bar pilots' equipment, which has often been one-of-a-kind in the world, reveals features uniquely designed to face the treachery of the bar and return all hands safe and sound to the sheltered inland environs of Astoria.

In the end, all of the stories send the same unequivocal message: the Columbia River entrance is no place to be without a bar pilot.

PART ONE

THE PROOF

CHAPTER 1

THE COLUMBIA'S UNIQUE GEOLOGIC HISTORY

WHEN IT COMES TO THE GREAT RIVERS of the world, the Columbia River does not spring immediately to mind. Rather, one envisions the mighty Amazon, Nile, Mississippi, or Yangtze. Why then is the Columbia one of the few places in the world where a ship entering a major waterway must take a pilot via helicopter rather than the more customary pilot boat? The reason is that the Columbia River outranks the planet's most famous rivers in navigational challenge, but not by the more common measures of overall length, volume, and current.

Without exception, the world's longest rivers present only modest challenge to the experienced ship captain who, aided by a local pilot, seeks to enter or depart one of these great rivers. This is because the Nile, Amazon, Mississippi, and Yangtze dissipate the energy of their tremendous flows through multiple branches across wide deltas. The mouths of the Nile and Amazon exceed 100 miles in width, while those of the Mississippi and Yangtze span more than 50 miles. In sharp contrast, the Columbia's mouth is constricted to barely four miles between a basalt headland and a hard sand spit guarding a deep Pacific Ocean trench that begins just a few miles offshore. This topographic combination so close to a deep sea trench has prevented a major delta from forming at the entrance to the Columbia River. Instead, there is a treacherous bar that generates the largest winter waves in the world.

The Nile and the Mississippi, for example, are mature rivers that show their age at their mouths in the form of wide deltas resulting from thousands of years of sediment buildup. As the Nile and the Mississippi approach the sea, there is little downward slope, and without the rush of current to self-scour a channel, these rivers spread out and meander across a broad deltaic plain of accumulated sediment.

To the modern oceangoing freighter, neither of these rivers presents any significant difficulty. With the ocean's swells running east to west, the north-flowing Nile and south-flowing Mississippi do not collide with a forceful wave condition. Both rivers have well-defined dredged channels that are easily transited under the guidance of a local pilot.

By almost every measure, the Amazon is the world's mightiest river. Second only to the Nile in length, the Amazon dwarfs its competition in every other category. The Amazon is the only river to draw water from two hemispheres. Its mammoth rate of discharge of five million cubic feet per second is twelve times the rate of the Mississippi and accounts for 20% of all the water that rivers send to the sea each year. Draining more than one-third of South America, the Amazon's basin is more than twice the size of its closest competitor, the Congo. The Amazon also has more miles of tributaries than any other river in the world.

But is the Amazon tough to navigate? At its mouth, the Amazon is so wide (250 miles) and so deep (120 feet) that entry is more akin to navigating into a large, deep, sheltered harbor than a river. It is little wonder that Portuguese explorers dubbed the confluence of a major tributary, the Rio Negro, with the main stem of the Amazon some 1,000 miles upstream near the port city of Manaus the "Rio Mar" or ocean river. Indeed, ships drawing up to 30 feet can transit the Amazon a distance of 2,300 miles upstream.

The Columbia River, often called the "Great River of the West," discharges more water than any other river in North America with the exception of the Mississippi. Compared to the incomparable Amazon, the Columbia's 1,210-mile length is just over one-quarter as long and carries only 10% of the Amazon's average flow. But what distinguishes the Columbia River from the world's most famous rivers is the narrowness of her mouth: a scant two miles between two massive jetties compared to 250 miles for the Amazon and 50 to 100 miles for the Nile, Mississippi, and Yangtze. Nowhere does a river push so much water into the ocean through such a narrow entrance.

Where is the Columbia's wide delta or estuary? Incredibly, it was destroyed some 15,000 years ago by the catastrophic floods produced by the breakup of ice dams impounding glacial ice melt in huge lakes across Oregon, Washington, and Montana at the end of the last Ice Age. Throughout the Ice Age, so much of the earth's water was stored on land that sea level was 300 feet below what we know today. The Oregon/Washington coastline extended some 25 miles west of the current shoreline, forming a gently sloping coastal plain that reached to the edge of the continental shelf. Present-day Astoria was 40 miles from the Pacific Ocean.

During the Ice Age, which lasted two million years, the mouth of the Columbia likely resembled the classic broad triangular delta crossed by multiple slowing branches or distributaries. But the Columbia River delta of the Ice Age was devastated at the end of that era by erosive forces of such colossal proportion as to defy the imagination.

Beginning 15,000 years ago and occurring periodically for another 2,200 years, the Missoula Floods were truly cataclysmic events, generating flows so enormous that they exceeded by tenfold the combined flow of all the rivers on the earth today and 60 times the flow of the Amazon. Put another way, a river that now averages just 10% of the Amazon's massive flow was transformed into a riverine tsunami that was 600 times greater than the river of today. And that was not just a single flood event. There were 40 cataclysmic floods over 2,200 years with the greatest producing a wall of water, ice, rock, and mud that reached a height of 600 feet in the Columbia River Gorge and surged downstream at speeds of 30 to 50 miles per hour. This is equivalent to a convoy of 200 semi-trucks surging side-by-side and stacked 30 high down an Interstate 84 that is as wide as today's Columbia River Gorge. Every tree, patch of soil, and human inhabitant along the river would have been scoured out and swept away.

The most spectacular changes to the Columbia River from the Missoula Floods occurred in the Columbia River Gorge and at the river's mouth. In the Gorge, a muddy, rocky slurry shot through the narrow gap in the Cascades with such force that it carved a deep V-shaped gorge, which changed multiple tributaries into spectacular waterfalls dropping as much as 620 feet before entering the river. The illustration on the opposite page depicts the Columbia River of today in the Gorge near Multnomah Falls and how the frightful leading edge of the greatest Missoula Flood would have appeared to someone on the banks of the Columbia River at that location. With individual Missoula Floods separated by many decades or hundreds of years, there is no doubt that native tribes resettled the river valley with its abundant game and fish in between floods.

The fact that Native Americans experienced the catastrophic Missoula Floods eons ago is corroborated by an Indian legend. A huge and ravenous beaver named Wishpoosh dominated Lake Kichelos at the summit of the Cascade Mountains. Coyote, the king of all animals, decided to kill the dreaded Wishpoosh and fought the huge beaver at Lake Kichelos. The battle was so fero-

MISSOULA FLOODS

The cataclysmic Missoula Floods of 13,000 to 15,000 years ago generated huge flows of water, ice, rock, and mud that formed the Columbia River Gorge and dramatically altered the entrance to the Columbia River. This scene depicts one of the largest of approximately 40 floods that reached a height of 600 feet in the Columbia River Gorge near Multnomah Falls, racing downstream at speeds of 30 to 50 miles per hour.

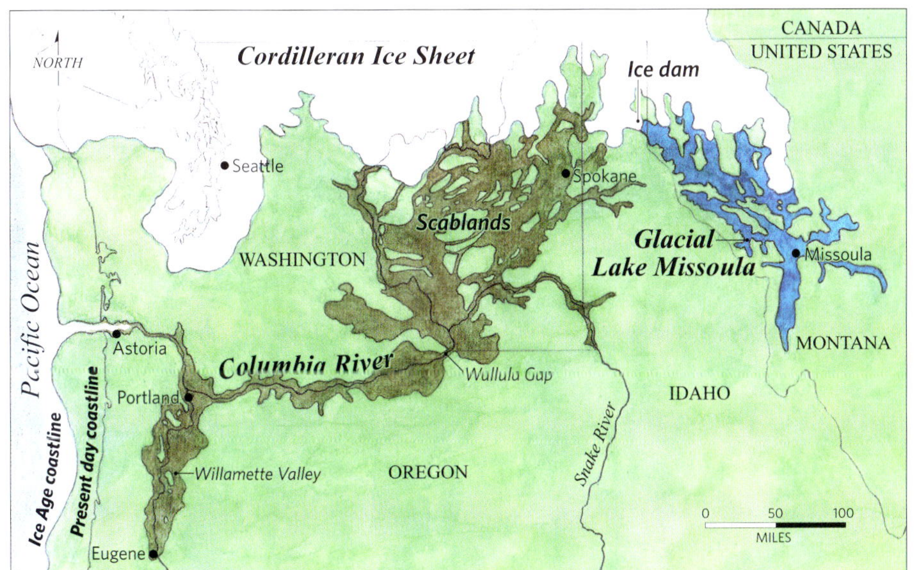

CORDILLERAN ICE SHEET

As the great ice sheet covering much of the Pacific Northwest gradually melted over 2,200 years at the end of the Ice Age, huge lakes were formed in Montana, Oregon, and Washington. These temporary lakes covered nearly 15,000 square miles. The massive floods occurred when ice dams created by the melting glacier broke up and the glacier retreated northward. With so much water stored on land during the Ice Age, the Pacific Ocean shoreline was 25 to 50 miles west of its present position.

cious that it whipped the lake into such a frenzy that its banks gave way and the water surged westward to form a new lake. The fight continued in the new lake and its banks gave way and formed another lake. This occurred repeatedly and as the epic fight moved westward with each new lake, a monstrous volume of water eventually flowed out to the sea along with the two animals.

Coyote then transformed himself into a stick that was swallowed by Wishpoosh. Now inside the beaver's stomach, Coyote used his knife to dismember Wishpoosh. Legend has it that he cast the pieces of Wishpoosh's body all over the land on either side of the Columbia and each one became one of the Columbia River tribes. And so, from the belly of a beaver came the Chinooks.

The awesome changes that the Missoula Floods wrought upon the landscape of the Pacific Northwest are most spectacularly evident in the Columbia River Gorge. It is here that basalt cliffs rise vertically to the sky and one sees the world's only example of a river cutting a pass directly through a major mountain range rather than meandering around the base of it.

One hundred river miles downstream at the mouth of the Columbia, the massive topographic change caused by the Missoula Floods is equal to the spectacle of the Gorge, but largely hidden from view. The 30-mile-wide

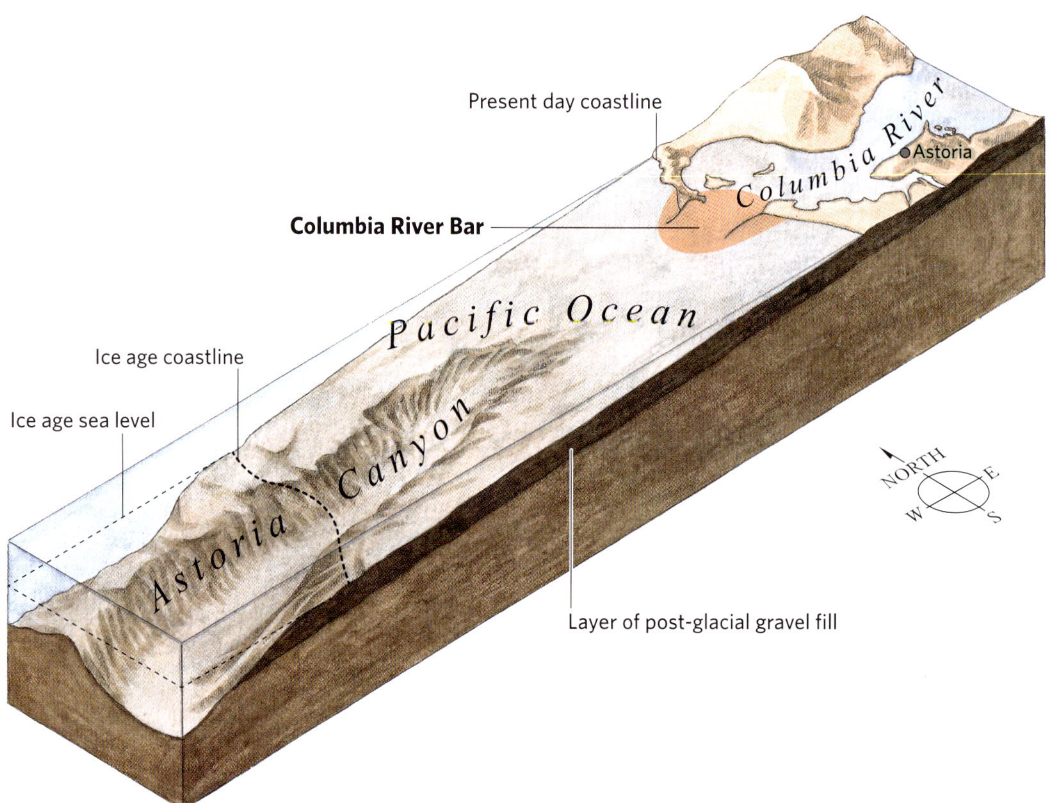

FOUNDATION OF THE COLUMBIA RIVER BAR

The Missoula Floods transformed the mouth of the Columbia River, eliminating an Ice Age delta, deepening the Astoria Canyon offshore and depositing huge volumes of post-glacial gravel fill. That fill became the foundation of the Columbia River Bar that can be characterized as a plume of shifting, finer sand sediments atop a hard gravel foundation.

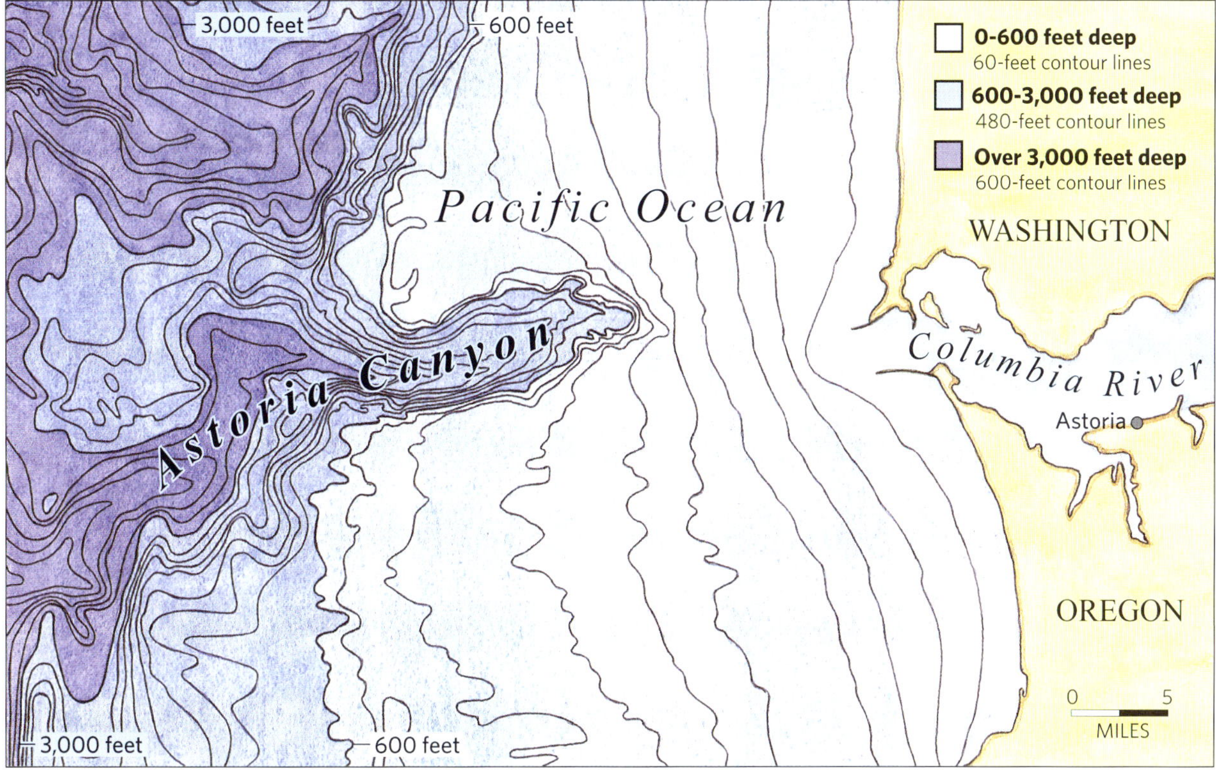

TRANSFORMING THE MOUTH OF THE COLUMBIA RIVER

This bathymetric chart shows how the Missoula Floods transformed the mouth of the Columbia River in a manner that prevents the formation of an energy-dissipating delta. The Columbia River carries fairly little sediment through a narrow entrance, just four miles between Cape Disappointment and Point Adams and only two miles between the ends of the jetties. The deep Astoria Canyon is only 15 miles offshore.

coastal plain nourished by the Columbia's Ice Age delta is gone, eliminated by the combination of a massive scouring out of the river's estuary and the swallowing up of the visible evidence by a rising ocean, except for the rocky headland of Cape Disappointment and the sandy Point Adams. In the place of fertile soil, the Missoula Floods deposited vast quantities of post-glacial gravel fill. Astoria now sits on 800 feet of this fill, some ten miles from the Pacific. In between Astoria and the deep water of the Astoria Canyon some 20 miles due west, nearly 1,000 feet of gravel fill deposited by the Missoula Floods established the foundation for what was to become the treacherous Columbia River Bar.

The incredible force and volume of the Missoula Floods also helped excavate, deepen, and lengthen the Astoria Canyon as shown above. Nowhere on the West Coast of North America is there a deeper canyon closer to a major waterway entrance.

In geologic time, events of 12,000 to 15,000 years ago are considered modern. But unlike the great rivers of the world, even with eons of time, the post-Ice Age Columbia River will not develop a broad delta. Why? Because there is no room. With only 15 miles of ocean between the river's mouth and the edge of Astoria Canyon, which drops from 300 to 3,000 feet in the space of 15 miles, there is room to develop a sandy deposit called the bar, but no room to dissipate the river's energy across a wide delta or estuary. The Columbia River Bar extends out from the pincher-like Clatsop and Peacock Spits in a plume that resembles an underwater delta. It is this bar that has presented such challenge to mariners over more than two centuries.

The origins of the Columbia River Bar are extraordinary, but what is the basis for the claim of "most dangerous?" Is there any scientific means of comparing the relative danger of entering or departing the world's major waterways? Yes and no. Today, the data is available to determine wave heights under various conditions at the entrances to any of the world's coastal inlets, but not so with respect to comparing conditions between waterways 50, 100, or 200 years ago. The best evidence from these bygone eras is found in the sailing directions published for mariners and in the anecdotal observations of explorers and ship captains.

Both types of evidence are worthy of careful examination because each is the best evidence in a particular era. The Columbia River Bar falls into two roughly 100-year eras separated by a two-decade transition period devoted to jetty construction. Looking over a period of 200 years, waterway entrance danger is a function of two factors: conditions (wind, waves, current) and vessel (type, size, and capability).

During the age of sail that prevailed for centuries before Captain Robert Gray's discovery of the Columbia River in 1792 and continued throughout most of the 1800s, the Columbia River Bar was impassable much of the year. Careful examination of the British admiralty's sailing directions and anecdotal evidence provides strong support for characterizing the danger of the Columbia River Bar as the worst in the world during this period.

In the modern post-jetty era, however, the Columbia River entrance has been engineered just like the other great rivers of the world to have a navigable entrance channel that can be transited by the oceangoing vessels of the world in most conditions. In the modern era, determining which waterway entrance is the most dangerous can now be examined as a matter of science. The key question to be answered is: which waterway experiences the worst wave action? This is because, as with the sailing ships of yesteryear, it is tall waves that pose the greatest risk to the seagoing vessels of today.

Despite the benefits of modern navigation technology including radar, GPS, and autopilot systems, when bad weather roils the sea, the most important asset on a vessel's bridge is not equipment, but a skilled sea captain or pilot. Statistics maintained by the International Union of Marine Insurance tracking total losses of vessels larger than 500 gross tons show that an average of approximately 100 ships per year were lost in the five-year period 2006-2010. Fully two-thirds of these total losses were caused by weather or grounding, the two greatest risks during a transit of the Columbia River Bar.

CHAPTER 2

PRE-JETTY CONDITIONS: 1775-1893

OVER THE MORE THAN TWO CENTURIES since European explorers first entered the Columbia River, what became known as the fearsome Columbia River Bar has undergone considerable change. The bar first documented by Spaniard Bruno de Heceta in 1775 and then first crossed by America's Captain Robert Gray some 17 years later in 1792 aboard the river's namesake ship *Columbia Rediviva* was impassable much of the year.

The treacherous conditions at the river's mouth severely limited the days during which a bar crossing could be attempted for more than a century until the south jetty was completed in 1895. The 1886 edition of the British Admiralty's sailing directions for the North Pacific characterize the Columbia as "by far the most considerable" of any river to enter the sea on the east side of the Pacific Ocean. Sailing vessels are warned that, without the aid of a steam-powered tugboat, the "entrance to the Columbia is impracticable for two-thirds of the year." It is flatly stated that the bar cannot be crossed at night and during daylight only under particular tidal and wind conditions. "Unlike all known ports, it would require both the tide and the wind to be contrary to ensure any degree of safety."

Before steam tugs were introduced at the river's mouth in the mid-1800s, sailing ships had to lie at anchor in Baker Bay for weeks at a time in winter months for that combination of fine weather and fair wind deemed indispensable to a safe outbound passage. Even more difficult for the sailing vessel was entering the Columbia River in winter because the lack of shelter from the prevailing winds and the heavy swell impeded a ship's ability "to watch her opportunity so conveniently as when at anchor in Baker Bay."

The whole purpose of the Admiralty sailing directions is to provide ship captains with instructions on how to navigate the oceans of the world and to enter navigable rivers and bays. But the 1886 edition goes only so far as to give a "general description" of the Columbia River, because "directions for entering it would be worse than useless owing to their inevitable uncertainty."

This dire warning continues:

> *The shipmaster, therefore, who would enter unassisted by a proper pilot, must be quite prepared to exhaust all a seaman's energies in the formidable undertaking. And he will the more readily comprehend the necessity of this, when it is said that, previously to the establishment of the pilot service, nearly every vessel whose visit is recorded gives an account of her grounding on some part of the passage; and the* Peacock, *one of the United States' Exploring Expedition, was totally lost on the terrible bar.*

These 1886 directions characterize the Columbia River entrance as "safe as that of most other barred harbours," owing to the combination of powerful steam-tugs and the bar pilots. However, because of shifting channels, it is emphasized in italics that no ship should attempt to cross the bar without a bar pilot.

The Columbia River entrance discovered by Heceta in 1775 and named by Gray in 1792 was over six miles wide between the two jaw-like headlands of Cape Disappointment on the north and Point Adams on the south. Large unstable shoals of hard sand extended from each headland, Peacock Spit from the north and Clatsop Spit from the south. Just inside the two spits lay a dangerous triangular wedge of shoal known as Middle Sands, only a small portion of which, Sand Island, was visible at high water.

According to Charles Wilkes, commander of the U.S. Exploring Expedition, that surveyed the river entrance in 1841, this wedge-shaped shoal was situated just beyond the jaws of the two typically breaker-laden shoals. This created two channels, north and south, but only one was generally navigable. When Captain Gray entered the river for the first time in May 1792, the official log describes an entrance by way of the north channel. Gray's ship began her bar crossing a "little to the windward of the entrance" and then ran in "north-north-east between the breakers." The *Columbia Rediviva* anchored "about five leagues from the entrance," opposite an Indian village at a spot near Point Ellice, a future campsite of Lewis and Clark.

The discharge of a mighty river flowing around an oft-changing Middle Sands and then between two shifting breaker-laden spits for twice daily battles with strong tidal currents was a prescription for rapidly changing shoals, and rapidly changing channels.

It was likely fortuitous that Captain Gray made his return to the Columbia River Bar in May 1792. From October through April, breakers often obscured the river's entrance. These same breaking waves rendered entry impossible and raised doubts about the size and significance of the river. Just one year before in 1791, Captain Gray had been off the mouth of the Columbia and had tried unsuccessfully over nine days to cross the bar. But he remained convinced of the existence of a mighty river by the color and current of the river water he encountered at sea.

One year later, Captain Gray's conviction was proven out. It was a singular achievement that might have gone to another explorer were it not for Gray's perseverance and patience in waiting for favorable conditions. Spaniard Heceta made it nearly midway between Cape Disappointment and Point Adams, but turned back after encountering currents and eddies that raised fear of running aground on Peacock Spit "notwithstanding a press of sail." These currents and eddies caused Heceta "to believe that the place is the mouth of some great river, or of some passage to another sea" and he named it Rio San Roque.

In 1788, English merchant John Meares reported reaching the approximate latitude of Heceta's Rio San Roque, which was marked on Spanish charts. He observed breakers running clear across a large opening and later concluded, "We can now with safety assert that there is no such river as that of St. Roc exists, as laid down in the Spanish charts." His low opinion was reflected in his naming of the headland guarding the northern entrance, Cape Disappointment.

Even an explorer of the caliber of England's Captain George Vancouver was fooled by the unusual convergence of the Columbia River and the Pacific Ocean. When Vancouver was off the breaker-bound entrance to the Columbia in April 1792, little more than two weeks before Gray entered, he observed the capes described by Heceta and "river coloured water," but did not consider "this opening worthy of more attention" and sailed north.

Later that year in August 1792, after receiving a copy of Gray's sketch of the mouth of the Columbia, Vancouver sailed back to the Columbia River entrance aboard the 340-ton *Discovery* and in the company of the 135-ton *Chatham,* commanded by Lt. William Broughton. In the fair weather of August without the heavy swell and breakers, Vancouver noted "this opening in the coast to be much more extensive than I had previously imagined."

Vancouver sent Broughton's *Chatham* to transit the bar, which was successfully accomplished via the north channel. Broughton spent three weeks exploring the river's mouth, and his log contained a vivid description of the Columbia River Bar:

> *I have never felt more alarmed or frightened in my life, never having been in a situation in which I conceived there was so much danger. The channel was narrow, the water very shoal, and the tides running against the wind at the rate of 4 miles an hour raising a surf that broke entirely around us, and I am confident that in going in, we were twice the ship's length from the breakers, that had we struck on we must inevitably have gone to pieces without the most distant hope of a life being saved.*

Broughton's chart shows the depth of this channel at 24 feet at low water, but conditions had obviously changed by 1813 when the British sloop *Raccoon* struck the bar while attempting a crossing relying on the Broughton chart.

Sir Edward Belcher in the *H.M.S. Sulphur* performed the first complete survey of the bar in 1839. Belcher found two navigable channels and named the north channel after himself, Belcher's Channel, and the south after his sovereign, Queen's Channel. Lt. Charles Wilkes, surveying for the United States in 1841, also found two channels with the greatest depths on the bar of 25 to 30 feet. While standing off Cape Disappointment planning his entrance, Wilkes wrote: "I am at a loss to conceive how any doubt could ever have existed, that here was the mouth of the mighty river, whose existence was reported so long before the actual place of its discharge was known, or how the inquiring mind and talent of Vancouver could have allowed him to hesitate."

Wilkes also described one ever-present difficulty in crossing the bar, "cross-tides changing every half-hour." According to Wilkes, "These tides are at times so rapid, that it is impossible to steer a ship by her compass, or maintain her position, and no sailing directions can possibly embrace the various effects produced by them...the safest time to cross the bar is when both the tide and wind are adverse; and this is the only port, within my knowledge, where this is the case."

By 1850, the north channel had filled in considerably and was declared unsafe for large vessels while the south channel was becoming deeper. These findings were made by Lt. Commander William P. McArthur and Lt. Washington A. Bartlett in the first U.S. Coast Survey in 1849-50. Compared to the Wilkes chart, Sand Island had moved one mile west. In noting the dramatic alterations to the estuary in less than a decade, McArthur commented, "This change will go on until some violent storm will throw up the sand again, and upon subsiding leave the water of the river to find a new channel."

One can try to simulate the extraordinary dynamics of the pre-jetty Columbia River Bar by playing with a series of differently shaped triangular rocks in a creek or small stream just above and then below two rocks acting like the jaws of Peacock and Clatsop Spits. By changing the size and shape of the wedge-shaped rock simulating Middle Sands and its distance from the two spits, one can create an infinitely varied range of currents and eddies.

Why place the wedge rock both above and below the rock spits in this rudimentary hydrology experiment? Because it's a good way to simulate the effect of the twice-daily high and low tidal influences on the flows at the bar and around what was Middle Sands. When the tide is incoming or in flood stage, your wedge rock is upstream of the spit rocks to simulate the Pacific's salt water surging eastward into and around Middle Sands. But when the tide is outgoing or in ebb stage, your rock should be downstream of the spits to simulate the even faster currents produced when the tidal and river currents combine to flow in the same direction.

The force of the water on Middle Sands is therefore greater on the ebb than on the flood. It should be no surprise then that the most dangerous bar conditions occur during a strong ebb tide. This is not just a function of the stronger current. As the ocean recedes at low tide, the Columbia's huge discharge races faster toward the bar, creating a more forceful collision with the incoming ocean swell. In winter, these swells typically approach the entrance directly from the west at heights of 10 to 20 feet. Once on the bar just outside and between the two spits, these swells steepen as they encounter the shallower water on the bar and then steepen further against the force of the outgoing fresh water, often producing huge breakers.

These were no doubt the conditions that greeted a nineteenth-century Italian missionary who spent six years traveling the West Coast in the mid-1800s. On his first trip to the Columbia River, Father Louis Rossi was aboard the steamer *Columbia* in 1856 on a voyage from San Francisco to Astoria when the vessel encountered a winter storm just offshore of the Columbia River entrance. The steamer spent a night and most of a day offshore trying to ride out the storm before attempting a bar crossing.

Father Rossi described "an enormous row of waves breaking for a distance of three leagues (nine nautical miles) from Cape Disappointment to Point Adams, and forming a kind of sandy crescent about 1,500 meters long at the river's mouth." Where the sea met the river, the "frightful impact" produced a roar so loud it could be heard several leagues away and he observed waves that reached heights of up to 60 feet.

Father Rossi was the first to categorize the Columbia River Bar as the world's worst:

> *Expert English, American and other navigators have asserted that there is no worse sea lane than this in the known world. Neither the English Channel, the Straits of Gibraltar, nor the Mexican Gulf can be compared with it. Its currents, rip tides, its storms, its sudden wind changes make it exceptionally dangerous, and the enormous bar does nothing to lessen the danger, especially in bad weather.*

Before the advent of steamers on the West Coast, the only vessels entering the Columbia were under sail. Between 1792 and 1875, 62 major wrecks occurred at the Columbia River entrance, which led to the bar securing the well-earned title "Graveyard of the Pacific."

CHAPTER 3

JETTY CONSTRUCTION: 1885-1917

"BAR IS BARRED," SHOUTED THE HEADLINE in the *Astoria Evening Budget* in 1929, reporting that the U.S. Geographic Board had finally agreed to change the reference to the mouth of the Columbia River on all maps and charts from the risky-sounding "bar" to the innocuous "entrance." The Astoria paper joined a chorus of news reports from around the Pacific Northwest claiming that two massive jetties at the river's mouth had eliminated the bar years earlier. The long campaign of Columbia River ports, the Portland and Astoria Chambers of Commerce and the Portland Steamship Operators Association had finally succeeded — the fearsome Columbia River Bar was no more!

But the real story is far different. Constructed over more than three decades between 1885 and 1917, the north and south jetties required nine million tons of stone to extend a combined total of nine miles into the Pacific Ocean. At the time, the Columbia's two jetties were the world's largest. Just the stone would fill 45,000 modern dump trucks.

While fairly remarkable feats of engineering at the time, the history of the bar's jetties is not one of unparalleled success, but rather a compelling story of constant and creative adaptation by the U.S. Army Corps of Engineers. It was originally believed that a four-mile south jetty would deliver a deep and dependable channel, not only displacing the shifting Middle Sands, but also establishing a self-scouring channel requiring no maintenance.

The theory of jetty construction into the sea at the entrance to a major river is to focus flow in a narrower but deeper channel that maintains itself through the scouring effect of the faster "jet" of water adjacent to the jetty.

In the century that followed the completion of the south jetty in 1895, the Corps extended the south jetty, built the north jetty, and tried to arrest the encroachment of Clatsop Spit into the channel by constructing a third jetty and a system of permeable dikes. And all this was supplemented by massive annual dredging that began in 1956.

In the 50-year period of 1956–2006, Corps of Engineers dredges pulled 220 million cubic yards of sand from the mouth of the Columbia River, enough to fill more than 18 million dump trucks. Put another way, if one considers that the dredged entrance channel covers three square miles (six miles long by one half mile wide and 48 to 55 feet deep), 220 million cubic yards would rise to a height of more than a seven-story building across every square foot of those three square miles. And where was all this material dumped? Fully 90 percent was deposited in ocean disposal sites. Years of accumulating this material in locations just offshore of the entrance channel has further complicated the hydrology of the Columbia River entrance, generating dozens of studies and considerable change in disposal site location and dredged material management.

Throughout this entire century of human intervention, the Corps of Engineers has never claimed that the Columbia River Bar was eliminated, let alone tamed. Construction of the jetties actually extended the bar further into the ocean, shifting its outer sea face by approximately two miles. This near doubling of the pre-

THE JETTY TRAMWAY

The Corps of Engineers used a double track tramway, steam locomotive, and flatbed dump cars to construct the north and south jetties. Laid on a foundation of brush mattresses, jetty stone weighing 10 to 18 tons apiece were built up to a height of 30 feet.

jetty bar of three miles west from Cape Disappointment to a full five miles only increased size of the most dangerous zone. Indeed, the Columbia River Bar is the the world's worst on three counts. The Columbia River Bar is the world's longest, generates the tallest waves in winter storms, and is the only coastal inlet among the world's top wave producers to include a dangerous turn.

Indeed, the greatest shortcoming of the huge jetty system at the mouth of the Columbia River is its failure to produce a due westerly channel alignment that enables either the inbound or outbound vessel to set up for and transit the bar without having to execute a turn. The last time that the channel ran due west and the Corps characterized entrance conditions as "splendid" was in 1925–26. Since then, neither a third jetty nor massive dredging have stopped the advance of Clatsop Spit into the channel, creating a bulge to the northwest and an "objectionable" channel alignment that continues to this day.

The discovery of gold in California in 1848 brought major change to the West. Between 1850 and 1900, the population of West Coast states of Oregon, Washington, and California grew dramatically. With this surge in population, trade increased substantially between West Coast ports but also with ports on the East Coast and abroad. As the population increased along the Columbia River, the navigational limits of the bar became an increasing regional liability. By the late 1800s, the new generation of cargo ships was three times the size of the vessels that typified the prior century and could carry seven to 10 times the cargo. With drafts of 20 to 26 feet, this modern fleet was disinclined to enter the Columbia River even in good conditions, because the channel depth in 1882 was just 19 to 20 feet at low water.

The Astoria Chamber of Commerce was one of many civic organizations urging Congress to improve the Columbia River entrance. The Chamber complained that the shoaling at the bar was causing some vessels to divert elsewhere and others to depart with less than full loads. The Chamber also argued that the draft restriction on the bar was giving foreign shipping interests an advantage because British iron ships drawing 20 to 22 feet could carry more cargo than wooden American vessels drawing 23 to 26 feet.

Congress responded in 1882 by ordering the Corps of Engineers to reconvene its Board of Engineers to draw up a permanent plan to improve the mouth of the river. The Board analyzed extensive land and hydrographic survey data and in September 1882 issued a report recommending a 4.5-mile-long low-tide south jetty. It was to extend from near Fort Stevens in a slightly convex curve northward to a point approximately three miles south of Cape Disappointment. The Board projected that the stone jetty would produce a self-maintaining channel of 30 feet at low water at a total cost of $3.7 million.

In 1884-85, work began on the massive south jetty project that was to consume a full decade. The first year was spent ordering materials, hiring workers, and constructing both shoreside facilities and an approach trestle from Fort Stevens to the starting point

GONDOLA CARS FULL OF JETTY STONE

Steam locomotive and fully laden gondola cars headed to dump their load of jetty stone. *Oregon Historical Society, No. 52068.*

THE SIZE OF JETTY STONE

One of the typical 18-ton jetty stones loaded aboard a gondola dump car.

or root of the jetty. From there, jetty construction proceeded in three-step increments: extend the double track tramway atop the trestle; build a jetty foundation of brush facines or mattresses to minimize erosion; and then place huge quarried stones weighing 10 to 18 tons apiece on either side and under the tramway to a height of 30 feet at the crest.

Tramway construction became increasingly challenging as it was extended out to sea. The sand was too hard for the piling, 55 to 70 feet long and up to 25 inches in diameter, to be driven into the seabed with a conventional piledriver. Instead, a specially designed piledriver fitted with a powerful pump first water-jetted a hole into the sand as deep as 22 feet and then the pile was hammered into position. Each bent or section of tramway was 16 feet in length with four piles per bent gradually progressing seaward a distance of 4.5 miles. The piles were braced transversely at the top with four six-inch by six-inch timbers, capped with foot-square timbers that were 22 feet in length and then linked with 32-foot-long, 12- by 16-inch stringers. The stringers were bolted to the timber caps with 26-inch long one-inch bolts. Then came the ties and narrow gauge rail in 14-foot sections and a two-foot wooden walkway in between the two sets of tracks. In good weather, 1,000 feet of double track tramway could be built per month.

Underneath the 22-foot wide tramway and extending out another 18 feet, a brush facine or mattress was made of fir or willow brush. Its purpose was to reduce erosion beneath the jetty stone by arresting the movement of stones into the underlying sand. Center and side mattresses were generally three feet in thickness and wired together to form a 40-foot wide foundation.

Atop the tramway, a steam locomotive and up to 30 gondolas or flatbed dump cars hauled the huge basalt stones barged from quarries as far as 130 miles away in the Columbia River Gorge. Dropping the jetty stones resulted in some breakage and a somewhat haphazard

JETTY PILE DRIVER

Modified pile driver used by U.S. Army Corps of Engineers to extend double track tramway during jetty construction. *Oregon Historical Society, No. 52069.*

arrangement. Broken stones and those that landed with their long axis perpendicular to the jetty rather than parallel to it rendered the jetty weaker than the Corps of Engineers' specifications and resulted in considerable periods of repair. In 1892, more time was devoted to repair work than to new construction.

In the first four years of the project, construction proceeded slowly and there was little effect on the condition of the channel. Following more rapid construction in 1889, there were positive effects on the bar and the channel depth increased to 20 feet. This scouring out of the channel progressed and a board of engineers was convened in 1893 to evaluate the progress so far and whether any changes were necessary to the original project specifications. In a report issued in May 1893, the board colorfully described the jetty as "a long, thin, narrow backbone of solid material resting upon a very doubtful foundation, against which the forces in action at the locality have accumulated large quantities of shifting sands. These shoals in turn have been able to break the force of the waves and protect the jetty from destruction. Jetty integrity and the permanence of the present favorable condition of the channel over the bar depend upon the amount of this sand that can be accumulated."

In the board's opinion, the jetty's long-term durability was dependent upon a continued accumulation of this sand. Unless the huge jetty stones were reinforced with massive quantities of sand, the jetty would be unable to withstand the force of winter waves from the Pacific Ocean. In the first of what became many engineering adaptations by the Corps of Engineers, the board recommended that four groins be constructed on the north side of the jetty and that the height of the jetty be increased to 30 feet.

All of these recommendations were accepted by the chief of engineers and the terminus of the four-and-one-half-mile south jetty was completed in 1895. At this point, the channel depth had improved to 30 feet and as much as 31 feet in many places. The deeper channel generated speedy economic returns to the region. Within two years, the total value of the cargoes crossing the bar more than doubled the annual average for the prior decade. A proud jetty worker living in project housing at Fort Stevens penned a poem that appeared in the Astoria newspaper on September 24, 1895 that celebrated the completion of the south jetty and included the following stanza:

When this jetty work was first begun, the people would relate,
It is another Government scheme, a few men for to stake,
It is as fine a piece of work, as stands beneath the sun,
There was eighteen feet of water where now there's thirty-one.

Construction of the south jetty also improved the channel alignment. In 1885, the channel had a bearing west of south, but as construction proceeded, it gradually swung around to the north and delivered an optimal, almost due west direction by 1895. But over the next seven years, the channel continued to swing north and its formerly ideal depth of 31 feet had declined to 22 feet by 1902. As channel conditions deteriorated, federal legislation in 1899 authorized a survey and report examining the desirability of "obtaining a channel 40 feet deep at lowest water."

A specially designated board of engineers took up the project, revisited the complex symbiotic hydrology at the river's mouth and in 1903 issued a report projecting that a 40-foot channel could be obtained in three steps: extend the south jetty due west by 2.5 miles; construct a north jetty from Cape Disappointment to a point two miles north of the outer end of the extended south jetty; and channel dredging.

Construction of the south jetty extension began in 1903 and was not completed until 1913. The exposure of the structure further offshore to massive wave action in winter storms destroyed thousands of feet of trestle each year. The winter of 1905-06 was particularly destructive, wiping out over five miles of trestle. District engineer Solomon Roessler observed that the work conditions had to be seen to be appreciated because "no language can adequately describe the fierceness of the onslaught." According to Roessler, it was "perhaps no exaggeration to say that there is no work in progress in the United States today at all comparable with this one in the difficulties, uncertainties and dangers that arise at every stage of its construction."

The original 4.5 miles of the south jetty required ten years of construction, but came in substantially under budget at $2 million when completed in 1895. The 2.5 mile extension proved so challenging that it took another decade and cost nearly $8 million.

As the Corps of Engineers predicted, completing the south jetty to a length of seven miles improved channel depth, reaching 36 to 37 feet, but this was short of the specified project depth of 40 feet. Construction of the north jetty began in the fall of 1913 and proceeded at a quicker pace in its more sheltered location extending out from Cape Disappointment. The north jetty was completed in May 1917 to a slightly extended length of 2.3 miles at a cost of $5 million.

When finished, construction of the two massive north and south jetties had required over three decades, nine million tons of stone, and $15 million. The Columbia River entrance now had the largest pair of jetties in the world and a channel through the fearsome Columbia River Bar that met the 40-foot specification.

When the north jetty was completed in 1917, the initial effects on the channel were positive. The south side between the jetties deepened quickly to 50 feet. But the depths on the outer bar – that nearly two miles between the jetty termini and deep water – were only 40 to 42 feet. In rough sea conditions, these depths meant large waves on the outer bar and major risk to ocean-going vessels.

Over time, however, the combination of the two jetties appeared to be working. The channel depths through the outer bar were gradually increasing and by 1925 the channel was one-half mile wide with depths of 46 to 48 feet. But the jetties were also dramatically changing the sand distribution regime at the mouth. Both Peacock Spit and Clatsop Spit grew substantially, extending their shallow shoal waters westward and effectively increasing the northern and southern reaches of the outer bar. The sand displaced by the jetties to create the entrance channel was effectively adding to the western edge of the underwater bar and extending westward into the Pacific Ocean.

The Corps of Engineers was thrilled with the channel conditions in 1925. The "splendid entrance conditions" that year were produced by the "combination of deep water on the outer bar, excellent alignment of the entrance channel, and protected from westerly waves afforded by shoal depths on the western portion of the bar." But the "splendid" best-ever channel conditions of 1925 were not to last.

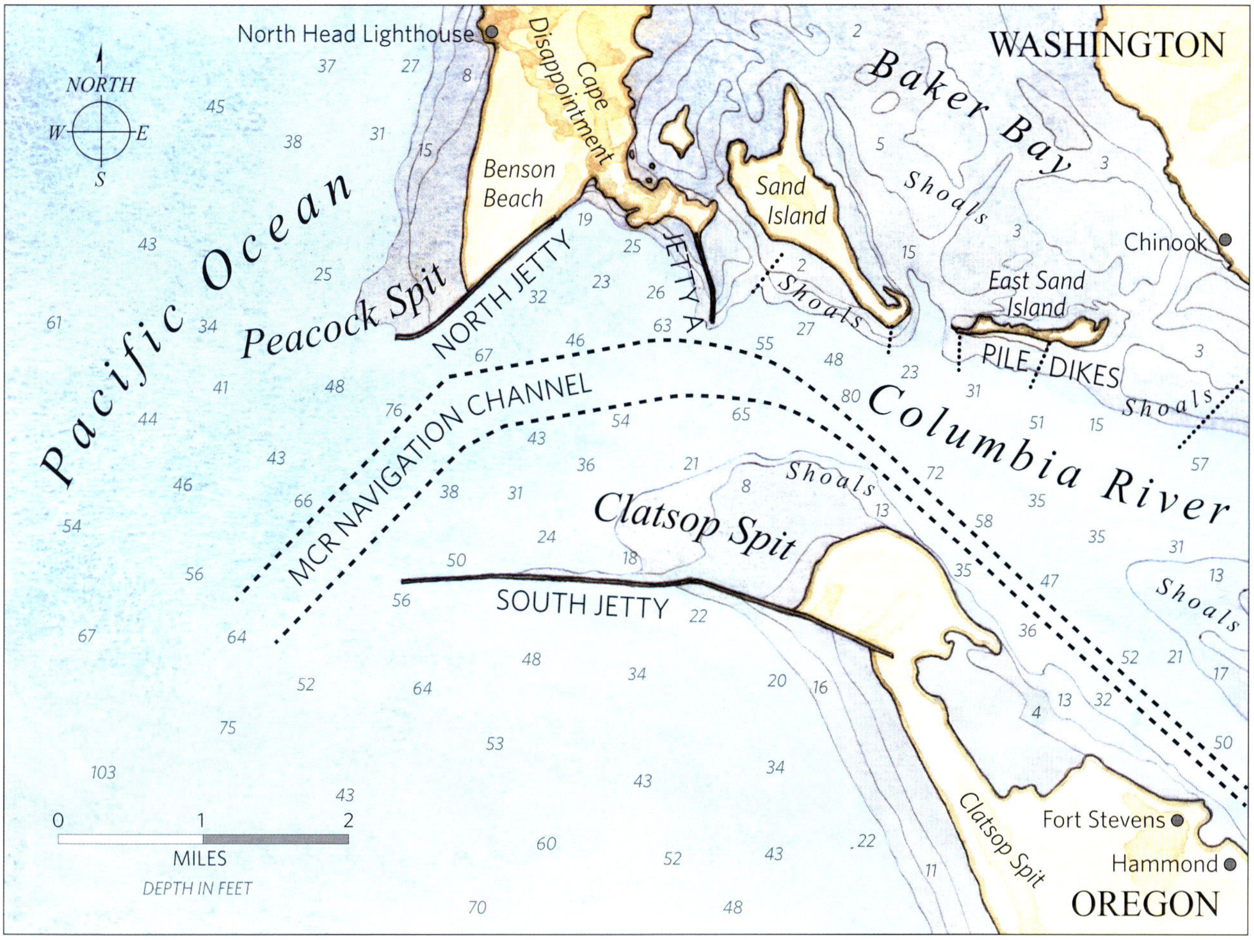

North Head Lighthouse
Cape Disappointment
WASHINGTON
NORTH
W E S
Baker Bay
Benson Beach
Sand Island
Shoals
Chinook
Pacific Ocean
Peacock Spit
NORTH JETTY
JETTY A
East Sand Island
PILE DIKES
MCR NAVIGATION CHANNEL
Columbia River
Clatsop Spit
SOUTH JETTY
0 1 2
MILES
DEPTH IN FEET
Fort Stevens
Hammond
OREGON

COLUMBIA RIVER ENTRANCE — PRESENT DAY

This current chart of the mouth of the Columbia River shows the configuration of the navigation channel and its pronounced dogleg turn.

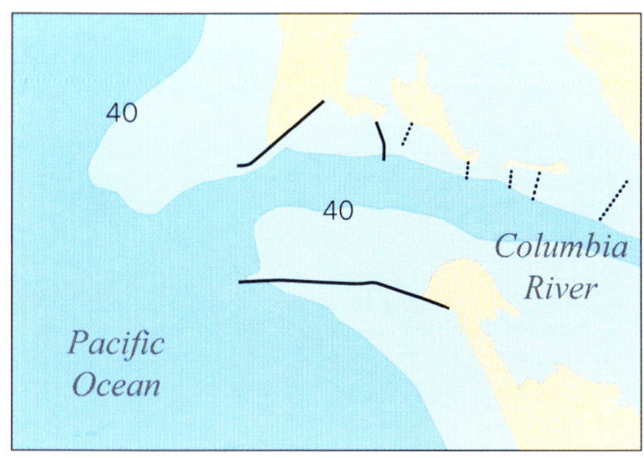

DEPTH CONTOUR — 1993

This 1993 chart shows that the shoal water of Peacock Spit and Clatsop Spit is receding and the 40-foot contour has retreated westward. This is the result of a combination of severe damage to the outer ends of both jetties that reduced their length by 5,700 feet, and winter storms.

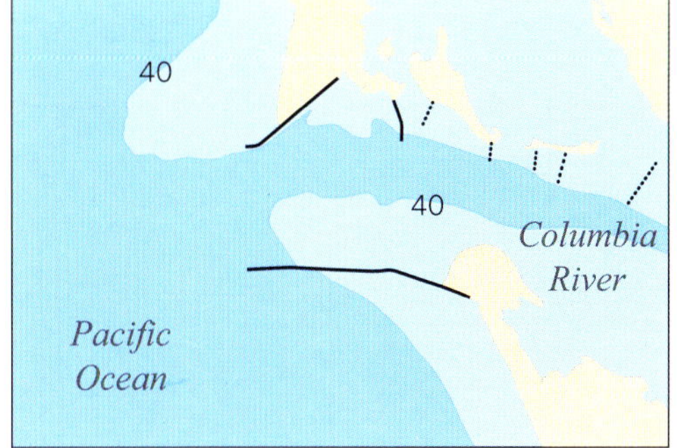

DEPTH CONTOUR — 2000

This 2000 chart shows the continued impact of winter storms and jetty degradation. Peacock Spit and Clatsop Spit continue to erode, receding faster in 1993-2000 than in the prior 60 years.

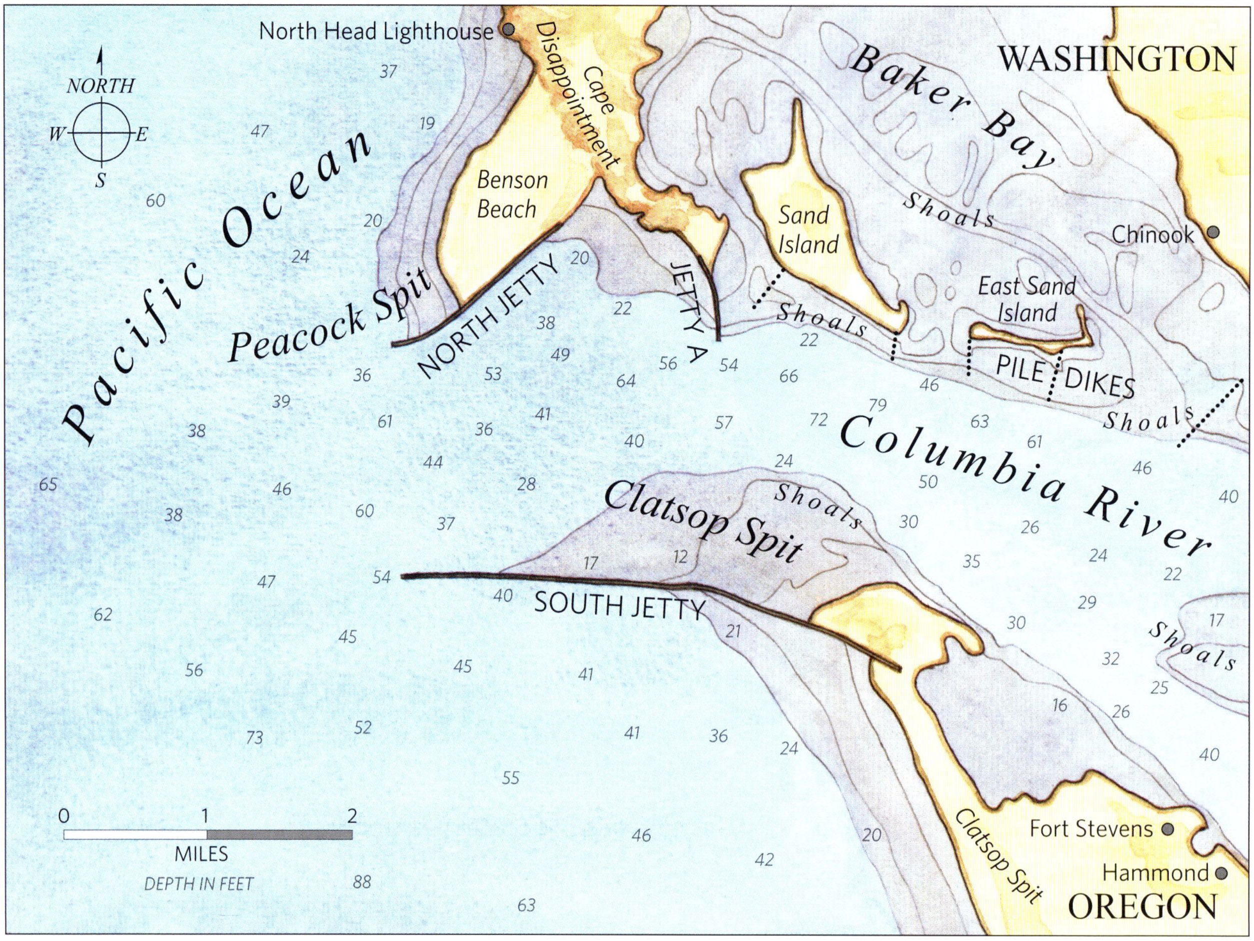

COLUMBIA RIVER ENTRANCE — 1950

After the north jetty and south jetty extension were completed in 1917, the navigation channel developed "splendid" conditions, running due west and deepening each year through 1925. But accretion on Clatsop Spit expanded its shoal northward and pushed the channel north as can be seen in this 1950 chart. This was the start of the dogleg turn that complicated transit of the Columbia River Bar.

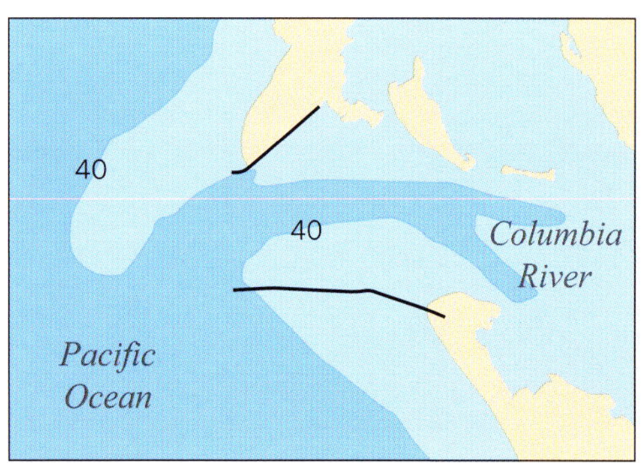

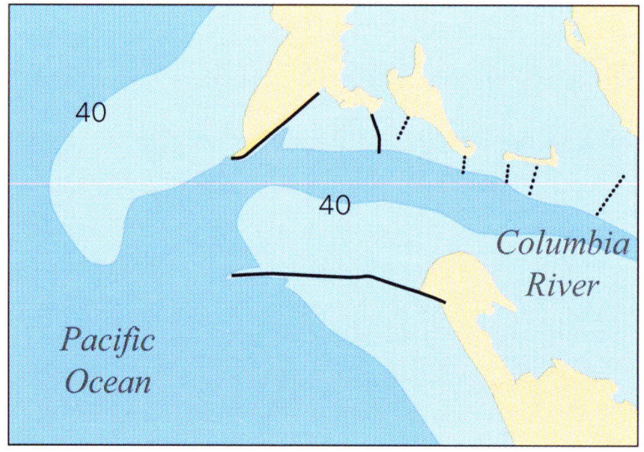

DEPTH CONTOUR — 1930

This 1930 chart shows the optimal channel conditions at that time: a 40-foot or deeper channel and a favorable east-west alignment.

DEPTH CONTOUR — 1950

This 1950 chart shows how accretion on Clatsop Spit and Peacock Spit is forcing reconfiguration of the channel to include a dogleg turn.

COLUMBIA RIVER ENTRANCE — 1839

This chart shows the Columbia River entrance in 1839 with the triangular shifting Middle Sands and the circuitous north and south channels around the shoals that made a bar passage impossible much of the year. These conditions existed from before Captain Robert Gray's discovery of the Columbia River in 1792 until the south jetty was constructed in 1885-1895.

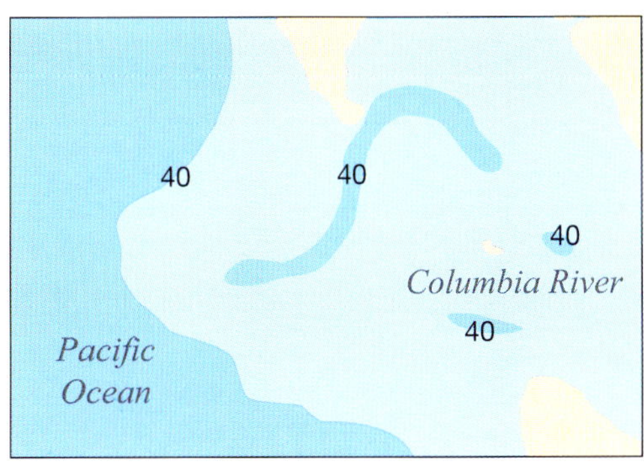

DEPTH CONTOUR — 1839

This illustration and the five others on this trifold use dark blue to show the 40-foot contour where the water depth is 40 feet or deeper. This illustration depicts conditions in 1839 when the mouth of the Columbia River was completely natural and Middle Sands continued to force vessels to transit a north or south channel around its shifting shoals.

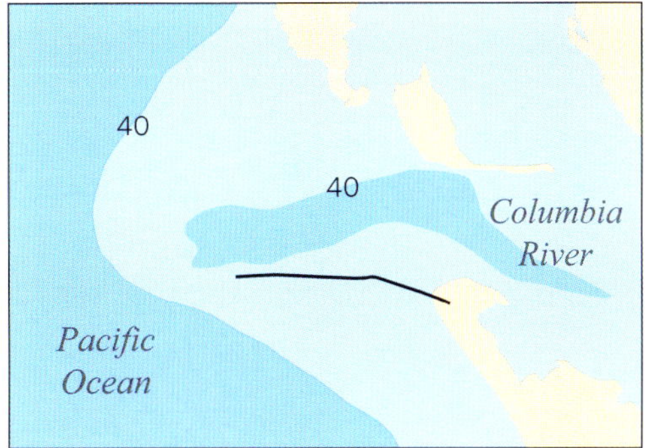

DEPTH CONTOUR — 1915

This 1915 chart shows the effects of completion of the south jetty. A 40-foot channel has been pushed through Middle Sands and extends out to sea to within one-half mile of the 40-foot contour in the open ocean.

CHAPTER 4

THE POST-JETTY BAR

FROM A NAVIGATION STANDPOINT, the 1920s were the golden years of the Columbia River entrance. The channel ran due west and was deepening each year, ultimately reaching depths of 46 to 48 feet, sufficient to handle any of the world's freighters of that era. With a width of one-half mile and due westerly alignment, the channel penetrated a bar that had grown sufficiently to the west to intercept and dissipate incoming storm waves. This dramatically increased the range of winter conditions where a bar pilot could thread the needle with an incoming or outgoing ship, sailing straight into or out of an entrance channel protected from large waves by shoal waters to the north and south.

Armed with such positive reports from the Corps of Engineers and the shipping industry, the campaign by the Chambers of Commerce in Portland and Astoria to strike the name Columbia River Bar from U.S. navigational charts gained traction. And in 1929, the U.S. Geographic Names Board agreed to eliminate the reference to a bar and changed its designation to Columbia River Entrance.

But this golden age did not last into the 1930s. The accretion of sand displaced by the jetties at Clatsop Spit continued and the spit advanced into the river channel and forced it northward. At the same time, Peacock Spit's low-water line was receding back close to the base of the cliffs of Cape Disappointment and the protective shoal waters of the outer bar were eroding away as well. The northward expansion of Clatsop Spit pushed that section of the channel lying inside the jetties to the north, transforming an optimal east-west alignment into a dogleg turn that the Corps of Engineers characterized as an "objectionable" alignment.

During the 1930s, the accretion of Clatsop Spit into the navigation channel continued, and the erosion on the north culminated with Sand Island being cut in two. To prevent additional northward migration of the channel and continued erosion of Sand Island and Peacock Spit, the Corps developed a plan to build a third jetty. Jetty A was to be constructed southward from Cape Disappointment along with four permeable dikes extending out from the south side of Sand Island. These structures were built during the period 1932 to 1938 and the south jetty was repaired and improved.

Over the next two decades, Jetty A and the dikes slowed, but did not stop, the northward migration of the channel. This forced the Corps in 1939 to begin dredging the northwest portion of Clatsop Spit to maintain the alignment of the channel. The Corps' hope for a return to a favorable east-west channel alignment had failed. The dogleg turn in the midst of a dangerous bar crossing had become a permanent feature.

In 1956, the entrance channel was deepened by dredging to 48 feet to accommodate the growing size of the world's fleet. That year saw the removal of over 14 million cubic yards of sand, more than the prior 10 years of maintenance dredging combined.

At the time, the upriver channel to over 100 berths at the Columbia's multiple ports was dredged to 35 feet, enough to accommodate ships drawing up to 33 feet with two feet of under-keel clearance and several feet more if the voyage was timed to take advantage of a tidal influence that was felt as far away as 100 miles upriver at Portland, Oregon, and Vancouver, Washington. As ships grew larger, Congress responded to calls for a deeper channel, authorizing a deepening of the river channel from 35 to 40 feet and its widening from 500 to 600 feet. This channel deepening project was completed by 1976. Curiously, there was no concurrent deepening of the entrance channel across the Columbia River Bar.

Uncomfortable with the idea that vessels drawing 40 feet could cross the bar with 48 feet of water in rough conditions, the bar pilots established a 38-foot limit to minimize the chances of hitting bottom while crossing the bar. According to the Corps of Engineers, this created a problem where the entrance channel was not "fully compatible" with the upriver channel. Studies ensued as a result.

Most illuminating was a ship motion study designed to measure heave (vertical acceleration), pitch (sideward acceleration), roll, heading, and position, but most importantly maximum vessel penetration – in other words, how deep the keel of a ship dropped below her draft. This study was conducted over three years and tested the four major categories of ships calling on the Columbia River: bulk carriers, oil carriers, container ships, and auto carriers.

The study confirmed what every experienced ship captain knows: "The dominant factor affecting vessel motions, and hence safety" is wave conditions. Wind, current, and weather conditions present complicating factors, but the dominant risk on a bar transit is wave action, which can cause disaster in any one of two ways. The first and most likely is a violent collision with the hard sand of the bar that could disable the vessel's steering, break her in two, or both. The other is broaching, getting caught sideways to the waves, capsizing and sinking or running aground and breaking up.

The 1978–80 study vindicated the bar pilots' concern that at least 10 feet of clearance with a 38-foot ship was needed to cross a bar with 48 feet of water. Of 53 vessels in the study, 15 percent had plunged from 14.3 to 26.3 feet below their drafts. The oil carrier *Hillyer Brown* encountered 15- to 20-foot swells inbound and 12- to 18-foot swells outbound, and dove 24.3 feet to a depth of 51.2 feet coming in and plunged 26.3 feet to a depth of 53.9 feet going out. Fortunately, tidal conditions were such that the bar pilots still had four to five feet of under-keel clearance on these voyages in what were characterized as "difficult" conditions.

Some studies end up collecting dust on the shelf. But this one was substantiated by two bottom strikes in 1980 and 1982. The inbound *Nanfeng* struck bottom three times in October 1980, hitting the hard sand twice between buoys 4 and 6 and once near buoy 8. At the time, the tide was two feet above low water and significant wave height was measured at 14 feet. With the vessel drawing 36.3 feet, it could plunge 14 feet at the bottom of a wave, penetrate 50 feet of water in the channel and hit bottom.

A second bottom strike in 1982 was more serious. The *Euroasia Concorde* was drawing 36 feet at her stern on an outbound voyage in January 1982. About midnight, she was one-half mile north of buoy 4 in a driving rain storm when the vessel hit a running breaker estimated by the bar pilot at 35 feet in height. With 35 feet of maximum plunge in a 48-foot channel, the ship's stern struck hard and the steering was rendered inoperable. A deep sea tug ultimately rescued the stricken *Euroasia Concorde* and towed her to Portland for repairs.

Armed with the ship motion study and the two incidents, the Corps of Engineers issued a feasibility report and environmental impact statement in 1983 examining multiple approaches to deepening the entrance channel. Interestingly, the Corps considered fixing the "objectionable" channel alignment by rotating the outer entrance channel (beyond the jetty tips) one-quarter mile to the north. But this alternative was dropped from serious consideration because it would have significantly increased the volume of annual maintenance dredging. By this time, the Corps had concluded that the bar pilots had developed the skill to deal with the dogleg turn in all but the most difficult winter conditions.

The focus of the Corps' feasibility study was on the optimal depth for the entrance channel. Relying on the ship-motion study and public testimony from the bar pilots and industry, the Corps recommended a 55-foot entrance channel. A statistical analysis of the ship motion data showed that there was a 95 percent probability that a vessel drawing 36 feet (including trim) and encountering 10 foot waves on the bar would plunge no more than 16.5 feet, leaving two feet of minimum clearance and a half foot to spare within a 55-foot channel. Deeper draft vessels would require higher tides in order to make the bar crossing with the minimum underkeel clearance.

In another nod to cost considerations, the Corps recommended that only the northernmost 2,000 feet of the half-mile-wide channel be deepened to 55 feet with the southern 640 feet left at 48 feet. This dropped annual maintenance dredging by an estimated 1.7 million cubic yards.

If you were to ask the Corps of Engineers why these shortcuts – no realignment and leaving 640 feet at 48 feet – were taken, you would receive two answers: cost and the skill of the bar pilots who had successfully worked the dogleg turn for more than 50 years and who could probably work with 2,000 feet of 55-foot water instead of a full half mile.

Congress appropriated the funds for the 55-foot entrance channel deepening project in 1983, and the work was completed in 1984. The Corps' estimate of the volume of sand removal required for this project was impressive. The agency predicted that 9.9 million cubic yards would be removed. The actual dredging totaled 8.9 million cubic yards.

Since 1984, the Corps of Engineers has dredged an average of four million cubic yards per year to maintain the entrance channel at its project depths. Why so much? Every year, the winter's storms send the Pacific's salty wedge deeper and with greater fury into the mouth of the Columbia than occurs throughout the other three seasons of the year. The result is a new winter deposit of Missoula Flood-generated glacial sand in the entrance channel, which then necessitates the maintenance dredging.

These same winter storms have substantially degraded the two jetties. The south jetty has lost 4,000 feet in length and there is severe damage in three locations along its trunk. Without repair, there is serious risk of a potential breach of the south jetty in a severe storm. The north jetty is also now compromised along much of its length. It has lost 1,700 feet in length, has a severely damaged head, and severe trunk damage with risk of breach in two locations. Even Jetty A, in its location more than two miles west of the existing ends of the north and south jetties, has suffered head loss and trunk damage.

As the ferocious environment at the mouth of the Columbia River has degraded and shortened the jetties, this in turn has contributed to a fairly dramatic erosion of Peacock and Clatsop Spits, the tidal shoals on which the two jetties were constructed and on which their long term stability depends. The continued degradation of the jetties is accelerating that erosion. Between 1993 and 2000, the 40-foot contour, where the depth of the Pacific Ocean reaches 40 feet offshore, has receded landward at a rate seven times faster than occurred during the prior six decades between 1930 and 1990. The loss of these shoals also makes the river entrance more exposed to larger swells and in turn higher waves because substantial wave energy is no longer lost on what were extended shoal areas. The result is greater risk of the type of swell and wave conditions that can broach even the largest cargo ship.

The Corps of Engineers made temporary repairs to the north jetty in 2005 and the south jetty in 2007 at a total cost of $27.5 million. Those temporary repairs are failing and the Corps in 2010 announced a 20-year major repair plan estimated to cost $400 to $500 million. The plan calls for capping the ends of both jetties with huge armor stones to prevent further erosion of the jetty heads. The agency also wants to build a series of groins to deflect the waves and currents that are undercutting the sand shoal foundation of the jetties. The groins will be added to the north and south jetties and to jetty A.

When it comes to navigational challenge, the presence of a bar at the entrance to a major waterway dramatically increases the treachery of entering or departing that coastal inlet. The "bar effect" increases danger in two ways. First, a bar increases swell or wave height. And second, a bar may cause waves to break so violently that a vessel's steerage is compromised, not in the open ocean where it may not matter, but in a narrow entry zone where steering clear of shoals is critical.

Fisherman and author John Paul Barrett was no doubt correct when he opined that the most likely place for a mariner to meet disaster is on a bar. Statistically, far more fishing and recreational vessels are lost trying to cross a bar than are lost in the open sea. As Barrett so aptly described, a "bar crossing is the roughest passage on the planet."

Rachel Carson, founder of the contemporary environment movement, vividly described the collision of wave and tidal current in her 1951 book *The Sea Around Us,* "Some of the most terrible furies of the sea are unleashed when tidal currents cross the path of waves or move in direct opposition to them."

But while anecdotal evidence is compelling, it lacks the force of scientific data that facilitates comparison of waterways on an apples-to-apples basis. In today's world, major waterways are all charted and there is comprehensive data showing ocean swell, period between swells, current, and significant wave height. Candidate waterway entrances were identified and provided to Professor Tuba Ozkan-Haller of Oregon State University's College of Oceanic and Atmospheric Sciences. Utilizing her expertise in marine geology, geophysics, and coastal engineering, Professor Ozkan-Haller assembled the relevant data and compared these major waterways in terms of probability of producing that most feared danger factor: high waves. This was accomplished with a mathematical equation that utilizes water depth at the entrance, speed of ebb current, period between swells and swell height in winter conditions.

Because the ebb tide colliding with the inbound ocean swell acts to steepen and increase wave height, the highest waves are encountered in ebb tidal conditions. The wave-amplifying force of these conditions over an entrance channel dredged to a given depth can be readily calculated. When the Columbia River Bar is compared to top candidates from around the world, it ranks first. When ocean swells reach a height of 19.6 feet offshore of the Columbia River Bar, the waves in the entrance channel will exceed heights of 31 feet. If the height of the offshore swell is 26 feet, which is not uncommon in winter, the wave heights will reach 42 feet.

WATERWAY	**WINTER WAVE HEIGHT**
Columbia River Bar	42.0
Port Phillip, Australia	35.4
Gironde, France	33.6
San Francisco Bay	28.7
Durban, South Africa	28.0

The second ranking waterway is Port Phillip, Australia, with wave heights of 35 feet. Gironde, France,

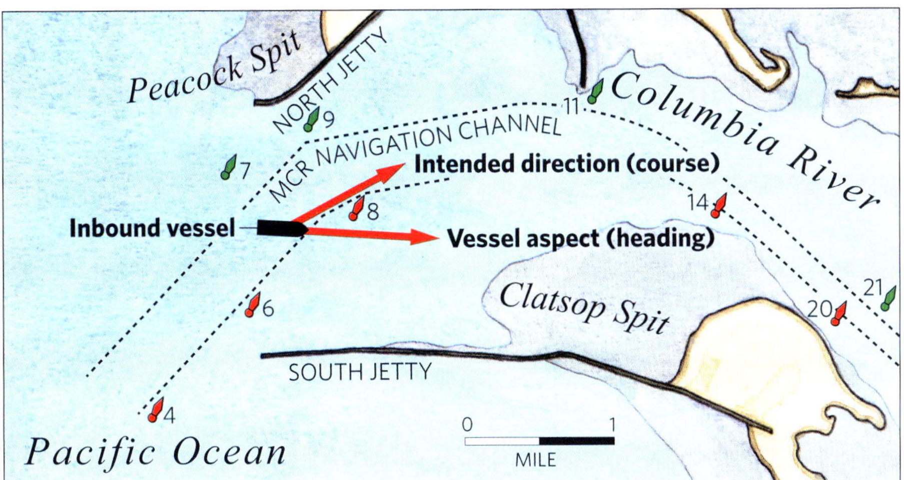

APPROACHING THE ENTRANCE

An inbound vessel fighting a southerly wind and northerly current must head on a course toward Clatsop Spit to stay in the channel and manage the dogleg turn.

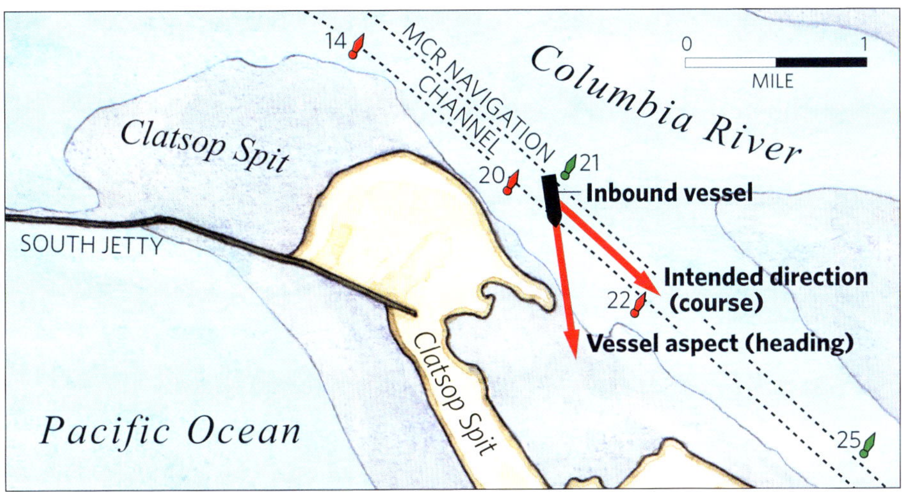

CRABBING TO STAY THE COURSE

Now past the dogleg turn, but still fighting the wind and the flood current, an inbound vessel must head toward Clatsop Spit to "crab" in a sideways fashion in order to maintain a course within the 600-foot navigation channel.

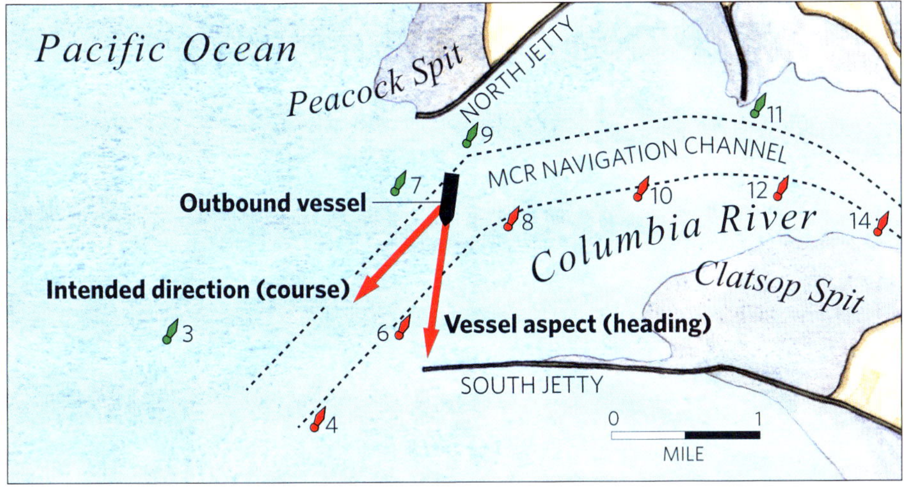

OUTBOUND TRAFFIC ALSO FACES CHALLENGES

An outbound vessel facing strong southerly winds and fighting the northerly set of the current must head straight for the south jetty to compensate for the conditions and stay in the channel.

is third at nearly 34 feet and San Francisco Bay and Durban, South Africa, are fourth and fifth at 29 and 28 feet, respectively. Remarkably, the Columbia River Bar is not only first in probable wave height during winter conditions, but is the only one of these waterways that includes a difficult turn during the transit through the entrance channel.

The impact of the dogleg turn in the middle of the Columbia River entrance channel is one of the bar's dangerous features that makes the presence of a bar pilot so critical to the safety of the vessel and her crew during a bar crossing. This is because winter conditions are commonly so severe that a ship must make maneuvers that her master would never perform, but if not performed in that circumstance, would result in a major catastrophe and cause huge environmental disaster and potentially the total loss of the ship.

During many winter situations, the wind from the south and current pushing to the north are so strong that the vessel, despite the intended course within the channel shown on the chart, must actually head on a course that is rotated more than 45° to the south from the intended course and repositioned in a way that it appears the vessel will pass on the wrong side of the buoy marking the southerly edge of the channel. As shown in the first illustration on the previous page, this radical maneuver must be made to compensate for the effect of the wind and the current. To the captain entering the Columbia for the first time or encountering these conditions for the first time, this maneuver appears to have the vessel headed for the shallows off Clatsop Spit, but in fact the vessel stays on course in the channel. However, unless this radical maneuver is used at the direction of an experienced bar pilot, the vessel will run aground on the hard sand of Peacock Spit or the jetty to the north.

In the middle illustration on the previous page, an inbound vessel is further along on its transit through the Columbia River Bar pilotage grounds and in the vicinity of Desdemona Shoals. The wind and the flood current are again setting the vessel to the north, pushing the vessel toward the shoals. To avoid running aground on the shoals and causing a major oil spill, the vessel must make the extreme maneuver of appearing to head directly toward Clatsop Spit in order to stay in the channel and avoid the shoals.

In a example involving an outbound ship (bottom illustration on the previous page), the vessel is attempting to stay in the channel with Peacock Spit to the north and Clatsop Spit to the southeast. Instead of pursuing a course with the bow headed parallel to the middle of the channel, the ship must be put on a heading that appears to head for rocks of the south jetty out of the channel to compensate for the extreme effects of wind and current.

But even this maneuver will not succeed in the face of 30- to 40-foot waves encountered between Buoys 6 and 8 where a ship must make a dogleg turn to stay in the navigation channel. In the last 20 years, several bar pilots have encountered waves of 30 feet or more, broached, and then spun completely around. In this perilous situation, these pilots knew their craft and the Columbia River Bar well enough to swing the rudder hard over, continue the storm-driven 360-degree pirouette, abandon the transit, and return to Astoria to await better conditions.

In sum, the radical maneuvers described and depicted in these three examples verify the bar pilots' long-held opinion that, unlike Houston or Baltimore or many other pilotage grounds, an incoming vessel cannot be "talked into" the protected inland waters inside the Columbia River Bar with radio instructions referencing navigational aids and communicating course changes. The maneuvers necessary to transit a difficult Columbia River Bar are simply too extreme to expect that a foreign master would ever execute the commands. It is vital that an experienced bar pilot be aboard to observe the conditions and deploy his or her experience in how best to maneuver the vessel to make that difficult crossing.

PART TWO

COLUMBIA RIVER BAR PILOTS: A BREED APART

CHAPTER 5

HIGHLIGHTS OF A PROUD HISTORY

TWO CHARACTERISTICS EXEMPLIFY the bar pilot tradition at the mouth of the Columbia River: ship captain's experience and courage in the face of danger; not the raw, undisciplined courage of the daredevil, but a courage born of years of seagoing experience and emboldened by the support of the world's best pilot transfer technology. While the bar has changed over the centuries, its treachery is timeless. The difficulty of boarding the inbound freighter in the rough winter conditions of today and then mastering wind, sea, and current through two jetties while setting up for a dog-leg turn no doubt equals the challenge of piloting vessels one-tenth the size of the modern freighter in the pre-jetty era of the 1800s.

The use of professional pilots dates back at least to Biblical times – the twenty-seventh chapter of Ezekiel refers four times to pilots. In classical Greek literature, the writings of Homer and Virgil contain multiple accounts of pilots. Homer's first book of the *Iliad* talks of the pilot Thesfor, "who guided the ships of the Achaeans to Ilium" and was "the best of augers, a man who knew the future, the present and the past."

During the fifteenth and sixteenth centuries, major maritime countries conferred high status on their top pilots and awarded titles such as Chief Pilot, Pilot Major, and Grand Pilot of the Realm. Pilots of this era not only took charge of a ship's navigation, but were expected to prepare charts of new waters and to collect navigational information to be incorporated into world charts and sailing directions for use by mariners throughout the world. For example, America's namesake, Italian explorer Amerigo Vespucci, was appointed Spain's Chief Pilot in 1508 with responsibility for the preparation of the charts to be used by other mariners.

In his capacity as Chief Pilot, Vespucci issued an order directing all who sailed into uncharted ports and harbors to make detailed reports to him so that charts could be prepared for use by other seamen. Sebastian Cabot, who explored much of the North Atlantic along the coast of North America, forwarded reports of his exploration to Vespucci in 1512. Great Britain's King Edward VI eventually appointed Cabot to the post of Grand Pilot of England, in 1549.

Over the centuries, a distinction developed between ocean pilots – those adept at navigating a vessel across oceans, and bar or bay pilots – those with expertise in local conditions with the skill to transit narrow straits and channels and avoid shoals. In modern times, the term "pilot" is used to describe the bar, harbor, or river pilot who boards the oceangoing vessel and brings specialized local knowledge to that part of a ship's voyage through often narrow and congested waterways. Practically speaking, the only way that a deeply laden ship can efficiently make such a passage without running aground is with the assistance of a skilled pilot. The specialized status of the pilot has led to universal recognition of the ship's "master" as the mariner or captain who pilots a vessel across oceans and the "pilot" as the chief navigator in rivers and harbors.

The term "pilot" is believed to have been derived from the old Dutch word *piloot*. According to Captain J. Edward O'Brien, the author of *History of Compulsory*

Pilotage, the term "signifies the steersman of a vessel, being equivalent to the Latin *gubernator* or *rector navis,"* but specifically "denotes one who works a vessel into port, or over a bar, or through a passage or channel."

In seventeenth-century America, each of the maritime colonies developed pilot organizations like those of their European counterparts, forming guilds or voluntary associations. These guilds employed apprenticeship programs to train new pilots, typically employing a set of rigid training rules over several years.

In the early colonial period, pilots were licensed without regard to the optimal number of positions needed to serve the port, and there was fierce competition between individuals or small groups of pilots. In a 1942 report on pilotage in America, the Commandant of the U.S. Coast Guard described this era as follows:

> *Up until about 1880 intense competition among the pilots prevailed at most ports, particularly those without a restriction on the number of pilots who might be licensed. Pilots cruised far out at sea in quest of business. Little effort was made to maintain pilot stations, and frequently pilots could not be had when wanted. In fact, the cut-throat competition among the pilots was unprofitable, wasteful, unsafe, and inefficient.*

Eventually, the colonies and then the individual states established pilot commissions to regulate pilotage rates. Competition between individuals or pilot groups in a given waterway gradually evolved into a single pilot association that pooled its resources to operate one set of pilot boats and generally assigned pilots to vessels in rotation.

When the U.S. Constitution was adopted in 1789, it granted to Congress the full power to regulate foreign commerce "with foreign Nations and among the several States," which clearly embraced pilotage. Recognizing, however, that the fledgling federal government was in no position to replace the existing state pilotage systems, one of the first acts of the First Congress of the United States was to ratify the state-regulated system. The Act of August 7, 1789, authorized continued regulation of U.S. pilots "in uniformity with the existing laws of the States respectively wherein such pilots may be, or with such laws as the States may respectively hereafter enact for the purpose."

The Congressional adoption in 1789 of the pilotage system in each of the states preceded by just three years Captain Robert Gray's discovery of the Columbia River. The earliest known pilot to serve the burgeoning fur trade in the early 1800s was Concomly, a one-eyed Chinook chief who commanded a large cedar dugout canoe powered by up to 20 Native Americans wielding cedar paddles. These canoes were fashioned from a single cedar log, reached lengths of 40 to 50 feet, and often were decorated with large carvings at the bow and stern. The canoe was literally "dug out" by burning, gouging, and carving the cavity in a process that could take one man up to three months.

Chief Concomly had a clear view of the river entrance from his lookout on Scarborough Hill behind Chinook Point. Once he sighted a ship, Concomly would often don a British soldier's red coat for the canoe trip to meet the vessel at a hailing point inside the Columbia River Bar. According to legend, Concomly was paid a fee of one axe and one blanket for piloting Hudson's Bay Company ships into Astoria. In his red coat, the Chinook chief was a striking figure with a flattened forehead that was a feature of the Chinook people of the lower Columbia. A Chinook baby's forehead was compressed with a cedar board for the first year of life to produce a steeply sloping forehead.

When Concomly died in 1830, he was buried near Fort George. Three years later, fort surgeon Meredith Gairdner excavated an Indian grave and found a flattened skull he believed to be that of Chief Concomly. It was sent to London as a medical curiosity and placed

TOWING ACROSS THE BAR

A small section of the starboard side of a steam tugboat operated by the bar pilots can be seen towing the French bark *Colonel de Villebois Mareuile* across the Columbia River Bar in approximately 1900.

in the Haslar Royal Naval Museum, where it was displayed for 114 years. Concomly's skull eventually made its way back home to Astoria in 1952 where it was displayed in the Clatsop County Museum for 20 years and then returned to the Chinook Tribe in 1972 and reburied in a secret location on the north shore of the Columbia River.

Unlike most other U.S. rivers and harbors, the mouth of the Columbia River did not experience a substantial period of unfettered competition between rival pilots. After taking over for Captain Jackson Hustler in 1852, the legendary Captain George Flavel dominated pilotage on the Columbia River Bar for more than three decades. Periodic efforts by others to enter the profession all failed within a year's time.

What was the Flavel secret to such remarkable early success? Flavel's singular achievement in building a strong and professional pilot organization was based on a powerful pair of practices that prevail to this day:

combining skilled ship masters with capable pilot transfer technology.

In his 35 years as "Czar of the Bar," George Flavel hired only experienced sea captains as bar pilots and built considerable loyalty within his customer base of ships trading between San Francisco and Astoria by often sending his pilots to San Francisco for their assignments, eliminating the need to sight and then board the incoming ship at sea.

Flavel was also no stranger to improvements in equipment. The pilot schooner *California* was only four years old when she began her service on the Columbia River Bar and Flavel was the first to introduce year-round tugboat service in 1869 with the launch of the 101-foot tug *Astoria* he had built in the same year according to his own specifications in San Francisco. In an era when sail was the predominant method of powering cargo ships, the steam towboat enabled the bar pilots to tow a sailing ship across the bar in conditions where there was little to no wind. The photo on the previous page shows what may have been the second tug *Astoria,* built by Asa Simpson and launched in 1894, in conditions that were likely at the rougher end of the spectrum in which this method was workable.

The massive construction project that built the north and south jetties between 1885 and 1917 produced a golden navigational era at the mouth of the Columbia River. The channel ran straight and true and the power of the hydraulic jet between the massive stoneworks gradually deepened the channel to unheard of depths of 46 to 48 feet. But in the 1930s, accreting sand at Clatsop Spit gradually forced a dogleg turn into what had been an optimal east-west channel alignment.

It is a major credit to the Columbia River Bar Pilots that the organization learned how to adapt to this unexpected development, which necessitated the radical ship maneuvers that are described in Chapter 4. It is notable that this new skill has saved the U.S. taxpayer many millions of dollars in dredging costs. In 1983, the Corps of Engineers elected not to eliminate the dogleg turn as part of its project to deepen the entrance channel from 48 to 55 feet because the bar pilots had developed the skill to negotiate that turn in all but the worst conditions.

As the U.S. economy expanded in the 1950s and 60s, led by growth in the western and southern states, there is little doubt that the Columbia/Snake River navigational system would have fallen behind its competition in Puget Sound and California were it not for the introduction in 1967 of the *Peacock* and her daughter boat boarding system. The drastic move to a self-righting version of a German rescue cruiser as the pilot boat to be stationed outside the bar in the Pacific Ocean with a motorized daughter boat likely ranks as the most significant event in the history of commercial pilotage on the Columbia River Bar.

Captain Kenneth McAlpin, who spearheaded the organization's review process that led to such a bold selection, deserves considerable credit. Under his leadership, the Columbia River Bar Pilots selected a pilot transfer technology that greatly expanded the envelope of their operating parameters to a level that was unknown to the pilotage world. With the *Peacock* and her daughter boat, the bar pilots could successfully board and debark ships in the ocean in swells of 20 to 25 feet. Today, as helicopter operations on the Columbia River Bar near the mid-point of a second decade, it is difficult to imagine how an entire organization of pilots was willing to embrace a revolutionary pilot transfer system and push it to its limits. The accomplishment came with multiple injuries and near deaths, but it was central to maintaining the Columbia River's reputation as a reliable waterway for international commerce.

Throughout the history of pilotage in the United States, nepotism has been a serious issue, with many pilot groups drawing a substantial percentage of their new pilots from the immediate relatives of the working pilots. Not so for the Columbia River Bar Pilots. In their first century, very few of the bar pilots had any family relationships, no doubt as a result of the danger and difficulty of the job and the historical emphasis on hiring only experienced ship's masters.

In their second century, the Columbia River Bar Pilots as an organization advocated that the ship's master tradition be made a legally required qualification for all applicants for new bar pilot positions. The requirement that all applicants have a minimum of two years' sea time serving as a ship's master or captain while operating under the unlimited license (any ocean and any ton vessel) has been a matter of regulation since at least 1940.

While a common requirement in other parts of the world, the Columbia River Bar is the only pilotage ground in the United States where the qualification standards are so stringent. As a result, the bar pilots of the last 50 years are drawn from throughout the United States after having graduated from a U.S. maritime academy and serving 15 to 25 years at sea, working up from an officer position to master. Although the modern organization of Columbia River Bar Pilots has an historic aversion to nepotism, there has never been an occasion in the last 30 years when there was ever a need to implement that policy because no son or daughter of an existing pilot has ever been an applicant for an open position in that timeframe.

The commendable commitment of the Columbia River Bar Pilots to a highly qualified and diverse organization reached its high-water mark when Captain Deborah Dempsey received her bar pilot's license in 1994. She was the first female maritime pilot in the history of the Pacific Northwest and, like her colleagues, had a distinguished career at sea before joining the bar pilots. A graduate of the Maine Maritime Academy, Captain Dempsey is a maritime woman of "firsts": first woman to command a merchant vessel in international waters and the first to skipper a merchant ship in wartime. Her 19-year career at sea included a harrowing episode where she was lowered by helicopter to a vessel drifting without power in a storm and then made emergency maneuvers to avert an environmental disaster and the loss of a $22 million ship. Her remarkable career is chronicled in her autobiography, *The Captain Is a Woman.*

From the standpoint of pilot transfer technology, the second most significant development in the 200-year history of the Columbia River Bar Pilots was the shift from the *Peacock*-centered station boat system to the helicopter/fast boat system in place today. The former performed well from 1967 until the late 1990s, but was being rendered increasingly unsafe and obsolete as the size of the average cargo ship more than doubled in the space of three decades. Given the entirely boat-based careers of every bar pilot, it took extraordinary leadership on the part of Captains Roger Nelson and Gary Lewin to persuade a majority of their fellow pilots to support the radical change to a helicopter as the primary means of boarding and debarking ships at sea, and generally being winched through the air to and from the ship's deck rather than landing on it.

The helicopter began operations in August 1999 and has safely transferred over 25,000 bar pilots to and from vessels in the Pacific Ocean for over 12 years without accident. The pioneering leadership of the Columbia River Bar Pilots in successfully introducing helicopter operations in the United States was recognized by *Professional Mariner* magazine with the award of its prestigious Plimsoll Award in the category of outstanding organization in 2003. As a member of the British Parliament from 1868-80, Samuel Plimsoll led the fight for legislation against unsafe maritime industry practices, especially the overloading of ships. His efforts were instrumental to the passage of the Unseaworthy Ships Bill in 1878 requiring ships to have marks showing safe load lines, commonly referred to as Plimsoll marks today.

It also took strong leadership on the part of Captains Wayne Stolz and Barry Barrett, who headed up the multi-year process of selecting the right type of pilot boat to pair with the helicopter. The fast, highly maneuverable, jet-powered *Chinook* and *Columbia* that operate today represented an untested choice in 1998, but both pilot boats have performed well since the *Chinook* was christened in 2000, and what became the fourth pilot boat *Columbia* was launched in 2008. And, in the decade since the *Chinook* arrived on the Columbia River Bar, no less than eight pilot organizations throughout the world have taken delivery of very similar pilot boats. With their studied but intrepid choice of new pilot boat technology, the Columbia River Bar Pilots have helped generate a world trend that is increasing navigational safety and efficiency.

At bottom, the core value underlying the 200-year history of professional bar pilots on the Columbia River Bar is navigational safety. While the *Peacock,* the helicopter, and the jet-powered pilot boats deserve top mention, a fairly modest but recent example helps prove this point. In the late 1990s, the U.S. Army Corps of Engineers gave notice that it was going to abandon the dredge ranges then serving as navigational aids to mark the edge of the dredged ship channel (some in water and others ashore). When the bar pilots raised questions about losing these navigational aids, the Corps offered to transfer ownership to the bar pilots. While some pilot groups in the same situation elected not to take over these navigational aids for fear of increasing their liability risk, the bar pilots accepted the Corps' offer and sought to secure the estimated $6,000 cost of annual maintenance in the 1998-99 rate proceeding.

In that proceeding, the Columbia River Steamship Operators Association successfully opposed the request, arguing that the bar pilots had failed to prove that the ranges were "indispensable" to safety. Despite that loss, the bar pilots refused to let the aids disappear, took title to the eight lower Columbia dredge ranges and resubmitted their request for the maintenance costs in the next rate proceeding two years later. The CRSOA continued to oppose funding for dredge range maintenance through the tariff, but this time the Oregon Board of Maritime Pilots approved the funding, finding that the bar pilots' "unilateral takeover and continuing maintenance" of the ranges demonstrated their importance to navigation and that the Board should defer to the bar pilots' "professional judgment in matters of safety."

Maintenance of the dredge ranges is a tiny item in a tariff that generates over $10 million annually, but this small bit of history helps prove a big point.

CHAPTER 6

THE TAI SHAN HAI: MODERN NEAR MISS

ON CHRISTMAS DAY IN 2002, the *Tai Shan Hai* departed San Francisco bound for Longview, Washington, some 40 miles upriver from the Columbia River entrance. A Chinese flag bulk carrier operated by Cosco Bulk Carrier Co., one of the world's largest shipping lines, the *Tai Shan Hai* was just shy of 100 feet in width and covered nearly two football fields in length from bow to stern. Now 16 years old, she was middle-sized by modern standards, but nonetheless was the equivalent of a 57-story building floating on its side and being pushed through the water by a 9,230 horsepower engine turning a single propeller.

The ship was scheduled to take on a cargo of petroleum coke for a return voyage to China where the petcoke would help fuel China's remarkable industrialization. The *Tai Shan Hai* sailed under a National Weather Service offshore forecast that warned of gale and possible storm conditions from northern California to southern Washington. Despite this forecast, Captain Guixin Tian elected not to fill all of his ballast tanks because he wanted the No. 3 hold ready to receive cargo upon arrival in Longview.

By early morning on December 27, the *Tai Shan Hai* was off the northern Oregon coast headed inbound for the mouth of the Columbia River. At the time, a storm warning and high surf advisory was in effect. The National Weather Service forecast predicted southerly winds of 50 knots with gusts to 65 knots and seas of 20 to 24 feet subsiding to 16 feet by late afternoon. The Columbia River Bar forecast was ominous, calling for seas of 18 feet with breakers expected during the ebb current. A subsequent weather update reported that seas off the southern Oregon coast had reached 23 feet and that the ocean swells near the mouth of the Columbia River would "build rapidly during the next few hours due to storm force wind associated with a strong Pacific storm system."

The ship's second officer first made contact with the Columbia River Bar Pilots at 0300 on December 27 when the vessel was approximately 35 nautical miles south-southwest of the Columbia River entrance buoy. The dispatcher advised him to call back when the ship was 15 miles from the entrance buoy. Meanwhile, Captain Charles Lane was piloting the Korean container ship *Hyundai Admiral,* which had dropped off containers in Portland and picked up others for its return voyage to the Far East. Knowing that the ETA for the *Tai Shan Hai* at the entrance buoy was 0600, Captain Lane planned to pilot the *Hyundai Admiral* from Astoria across the bar to the open sea where he would be hoisted off by helicopter and then transferred to the *Tai Shan Hai* for the inbound bar passage.

As the *Tai Shan Hai* neared the mouth of the river, Captain Tian observed "large waves and white caps" stretching across the river entrance and began to doubt that a bar pilot would actually bring his ship across the bar. He started to look for an anchorage area north of the entrance in case the bar was closed.

At 0500, the winter storm knocked out power at the Astoria Regional Airport, and the *Seahawk's* crew radioed that the helicopter was grounded due to the power outage. When this message was relayed to Captain

Lane, he quipped, "If I'm working in the dark, why can't the helo." Hearing the radio traffic that the helicopter likely would not be flying, the captain of the pilot boat *Columbia* self-dispatched to disembark Captain Lane from the *Hyundai Admiral.*

Captain Tian's decision not to fully ballast his vessel during a storm was now coming back to haunt him. Not filling the No. 3 hold with water caused the ship to ride higher in the water. As a result, as the *Tai Shan Hai* pitched fore and aft riding 20-foot plus waves, her propeller was coming out of the water and racing. This caused the main engine governor to cut in and temporarily reduce engine speed. Between 0530 and 0600, the main engine speed was reduced in an effort to deal with the pounding and racing of the propeller.

With the helicopter out of service, the bar pilot dispatcher radioed the *Tai Shan Hai* to stay five nautical miles offshore. Captain Tian's English was so poor, however, that he misunderstood and continued on toward the pilot station assuming a bar pilot would still board at 0600. By 0600, the *Tai Shan Hai* was just one mile south of the pilot station and on a course that would complicate Captain Lane's planned disembarkation from the *Hyundai Admiral* to the pilot boat *Columbia.* Debarking by pilot boat took an extra 30 to 45 minutes compared to a hoist off by helicopter. The ship had to turn into the wind to create a lee that provided calmer water to accommodate the pilot boat approach and enough time for the pilot to climb down the pilot ladder and then time his jump using one or two manropes (it's actually more like rappelling) to cover the last 10 to 15 feet between the side of the ship and the deck of the pilot boat.

Captain Lane called Captain Tian and requested that the *Tai Shan Hai* make room for his disembarkation by turning to starboard, passing south of the entrance buoy and moving 10 miles offshore to await further instructions. But the *Tai Shan Hai* kept on coming and repeated her request for a pilot. Captain Lane reiterated his instructions for the vessel to move out to sea, and the master and chief officer of the *Hyundai Admiral,* recognizing the problems with Captain Tian's language skills, intervened and tried to assist with the radio communications.

Captain Tian attempted to turn the *Tai Shan Hai* to starboard, but the strong winds, high seas, and reduced power from the continued uncovering of the propeller thwarted multiple attempts. As the ship sought to come to starboard, the wind and sea pushed her toward shallower water near the dredge spoil dump site just south of the river entrance. This was an area where the ship could encounter breaking seas and be forced aground on Clatsop Spit. Recognizing the danger of drifting into that area, Captain Lane called the *Tai Shan Hai* and urged Captain Tian to swing the ship to port. "The *Hyundai Admiral* is west-southwest of the entrance buoy. Head towards me and try to hit me," Lane urged.

For the next hour, the *Tai Shan Hai* tried without success to swing to port and move offshore. At 0700, Captain Tian radioed for tug assistance, but Captain Lane replied that conditions were too rough and it was time for the exercise of "captain's judgment" to remove the vessel from danger. As the *Tai Shan Hai* drifted into the Columbia River entrance channel, Captain Tian ordered his chief officer to ready both anchors.

At 0740, shortly following reports to the Coast Guard by Captains Lane and Tian of the developing emergency, the *Tai Shan Hai* dropped both anchors simultaneously in 66 feet of water. Biting into the bottom just north of the charted entrance channel, the anchors arrested the vessel's perilous drift toward Peacock Spit on the Washington shore. As the sun rose during the next 20 minutes, Captain Tian remained on the bridge as his ship battled southwest seas of 30 to 46 feet. The anchors appeared to be holding.

At 0750, with the *Hyundai Admiral* having made her lee, Captain Lane successfully disembarked to the pilot boat *Columbia*. But he had to abort the plan to board the *Tai Shan Hai*. Any attempt to board an anchored ship rolling heavily in 30-foot seas and sustained winds of 50 knots with no ability to create a lee for the pilot boat would have been suicidal. The tide was also beginning to ebb, increasing the risk of breakers in the ship's vicinity.

As the *Columbia* passed the *Tai Shan Hai,* Captain Lane observed that the anchor chains had adequate space between them, but were under such excessive strain as to be close to the breaking point. He also saw that the propeller was not turning. Captain Lane quickly called Captain Tian to suggest that he reduce the strain on the anchor chains by engaging his engine on a slow ahead bell and that a crewman be stationed on the bow to monitor the chains. Lane confirmed compliance with his suggestion when he saw a puff of smoke from the ship's stack.

While Captain Lane was busy advising the *Tai Shan Hai* and attempting to depart the *Hyundai Admiral,* Captain Mike Dillon, the bar pilot in rotation behind Captain Lane, had sprung into action. Informed of the *Tai Shan Hai's* plight, he proceeded directly to the airport to meet the helicopter crew and Coast Guard personnel. By 0800, the *Seahawk* had been returned to service after power at the airport was restored and an indicator light showing metal in the main rotor gearbox was traced down as a false alarm. In his meeting with the Coast Guard duty officer and others, Captain Dillon laid out his plan to fly out to the *Tai Shan Hai,* assess her situation and hold her at anchor if possible until the weather subsided. If her anchors did not hold, he hoped to somehow maneuver the vessel out to sea.

The *Seahawk,* an Italian-made Agusta twin engine helicopter, lifted off at 0915 with Captain Dillon and pilot John Glen and co-pilot Tom Thompson en route to the *Tai Shan Hai.* Ten minutes later, the helicopter was above the stricken ship that was rolling heavily as the seas washed broadside over her main deck. Despite never having transferred a pilot in 70-knot winds, Glen managed to set the *Seahawk* down on deck long enough for Captain Dillon to scramble out. As the helicopter lifted off, Glen recalled, "We took green water over the deck right where we had just landed." The helicopter briefly lost sight of Captain Dillon and hovered close by, caught sight of him entering the house and confirmed by radio that he was OK. Once in the wheelhouse, Captain Dillon assessed the situation, set up his GPS system and established communications with his office and the Coast Guard. He observed that the vessel was steaming at various speeds in an attempt to maintain a heading into the swell, but was at times rolling hard to starboard "with swell breaking on deck." Captain Tian appeared very concerned and well aware of his ship's precarious situation. Captain Dillon tried to bolster the Chinese captain's confidence with a quick recap of the pilot's previous experience waiting out severe storms in the Aleutians off Alaska when Dillon commanded Chevron oil tankers. "I told him that the winds were so strong that the radar antenna was unable to rotate, but we had complete success with anchors down steaming into the swell," recalled Dillon.

Captain Tian advised that both anchors were dropped at the same time, but had good spread between them. Doubting that this was possible with a simultaneous anchor drop, Captain Dillon went down to the forecastle to take a look at the anchors. This inspection, which was interrupted by a full soaking from a swell across the deck, confirmed his worst fears. The anchor chains led together into the sea with one or the other taking all the strain. Captain Dillon became concerned about the ability of the anchors to hold the ship and the risk of fouling both anchors. Their entanglement could cause one or both to part or to drag.

In fact, the anchors were dragging. In just over two hours, the *Tai Shan Hai* dragged anchor six-tenths of a nautical mile northeast of her original anchor position. As the chart below shows, the ship was now on the Washington side of the channel less than one mile from shallow water off the southern end of Peacock Spit. If she continued to drag anchor on her current heading, the *Tai Shan Hai* would run aground and break up within a matter of hours in the heavy swell.

While on the forecastle, Captain Dillon also assessed the crew that accompanied him from the bridge. The chief mate appeared both competent and able to understand the gravity of the situation. "I hoped that he had the ability, knowledge and conviction to slip the anchor should an emergency maneuver warrant," Dillon said later. But as he saw the boatswain and another crewman hiding under the fair-weather, Captain Dillon had less hope for their abilities to take quick and decisive action in an emergency.

For over three hours, Captain Dillon monitored the ship's position and the weather forecasts, occasionally running the engine at full speed to protect the anchors

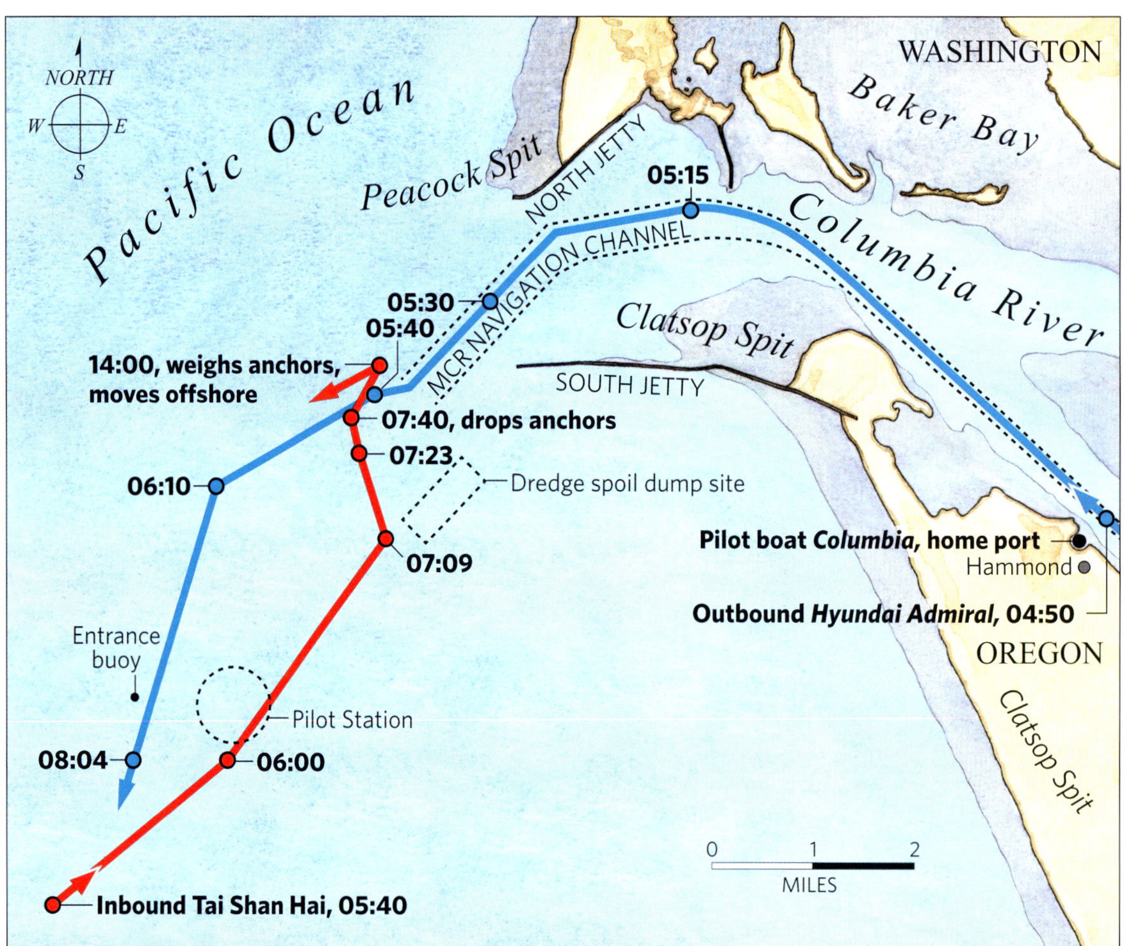

THE STORY UNFOLDS

This chart depicts the positions of the *Tai Shan Hai, Hyundai Admiral,* and pilot boat *Columbia* on December 27, 2002.

and mitigate the northward drag. When the forecast showed the storm front passing and the seas moderating by 1400, Captain Dillon made a plan to heave up the anchors if the forecast proved correct and proceed to seaward of the entrance buoy. He also cleared the plan with the Coast Guard.

By 1300, when there was a marked drop in sea and swell conditions, Captain Tian began to insist that the *Tai Shan Hai* weigh anchor and attempt to cross the bar rather than head out to sea. But Captain Dillon viewed the conditions as still marginal and refused. Fearing that the Chinese captain would continue to exercise poor seamanship and might attempt a bar passage without a pilot, Captain Dillon made the extraordinary request that the top Coast Guard official with jurisdiction over the Columbia River issue a Captain of the Port Order prohibiting the ship from crossing the bar without a bar pilot and only under moderate conditions.

When the conditions calmed further by 1400, Captain Dillon executed his departure plan. The anchors were successfully brought home, but even with orders for full maneuvering power, the *Tai Shan Hai* encountered difficulty staying on course and never generated the RPMs that would be expected at full power. It was only with considerable maneuvering that Captain Dillon was able to keep the ship from turning broadside to the incoming swell. In his view, these problems were the result of the ship's inadequate power plant for the conditions, the lack of proper ballast to keep the prop and rudder in the water, and poor communications between the deck and engine departments that continually resulted in lower levels of power to the propeller than ordered from the bridge.

Nearly two hours after heaving up the anchors, the *Tai Shan Hai* was safely offshore. Captain Dillon was hoisted off the ship's deck by the *Seahawk* at 1550 and landed at the airport at 1605. Before departing the ship, he confirmed that the Coast Guard Marine Safety Office in Portland had issued the Captain of the Port Order and he went over its terms with Captain Tian.

That order left nothing to chance. The loss of the ship could result in the spill of her nearly 200,000 gallons of fuel and a major environmental catastrophe. The *Tai Shan Hai* was directed to stay a minimum of 12 nautical miles offshore with her head into the prevailing seas and to maintain a communications watch with the Coast Guard. Because of her inadequate power, the ship was directed to remain offshore until entry approval for a bar passage was granted. To gain that approval, the order laid down four conditions: wind less than 20 knots; swells less than six feet; bar pilot boarding no closer than 12 nautical miles from land; and demonstrated maneuvering capability at sea speed satisfactory to the bar pilot.

The following day, December 28, bar pilot Captain Gary Lewin boarded the *Tai Shan Hai* via the *Seahawk*. While approaching the ship aboard the helicopter, he noted damage to the port side pilot ladder. Once in the wheelhouse, he checked out all of the vessel's navigational equipment and then conducted the propulsion and handling tests that the bar pilots had recommended and the Coast Guard had incorporated into its order.

Captain Lewin knew well how the *Tai Shan Hai* should handle if she was to be cleared to cross the bar that day. During a 15-year career at sea, including two years as a captain, Captain Lewin had experience in all sorts of weather all over the world. He'd even had the incredibly unique experience of delivering a tanker to Chittagong, Bangladesh, to be scrapped on a long Indian Ocean beach on literally his last job before joining the bar pilots in 1983.

Captain Lewin thought of that experience as he headed out to the *Tai Shan Hai* for her sea trials. He recalled that the tanker he captained was over 20 years old but of a similar size to the Chinese ship that had almost become involuntary scrap on Peacock Spit. After waiting through the night for the high tide, he maneuvered the ship north offshore of Chittagong and pushed the vessel to full ahead and maximum RPMs to a designated location where he waited for a radio message from the beach.

On the shore, a scrap yard manager monitored the ship's progress by following her lights and then radioed, "Turn." Captain Lewin ordered the helmsman to turn hard to starboard while keeping the engine full ahead. As the ship lumbered toward shore, he called out her heading every 60 seconds and adjusted course per instructions from shore. Several hundred yards from shore, Captain Lewin could make out the space on the beach his ship was supposed to fill between two derelict ships in varying stages of dismantlement.

The ship's bow generated a huge wake as the ship closed in on the beach, drowning out the engine. As her keel hit the sand, she continued to plow landward under the twin power of 25,000 tons of forward momentum and a single screw turning full speed until the ship's rudder hit bottom. The huge hull slid up the beach another 100 yards before skidding to a stop with a slight list to port. Captain Lewin ordered the anchors dropped, the engine and generators stopped and a rope ladder deployed to disembark his crew. Within hours, hundreds of ship breakers armed with torches were aboard, ripping the ship into pieces of steel piled by hand on the debris-strewn shore.

That Bangladeshi experience flashed through his mind as Captain Lewin readied the *Tai Shan Hai* for tests of her engine and maneuverability. In the relatively calm conditions prevailing that day, the ship was up to the task. He called the Coast Guard to report his findings, and the Captain of the Port order prohibiting her entry into the Columbia River was lifted. The *Tai Shan Hai* transited the bar during daylight hours on December 28, 2002, without further incident and made her way upriver to take on her cargo of pet-coke at Longview.

After disembarking the *Tai Shan Hai* in the protected inland waters of Astoria and handing her over to a river pilot for the balance of her voyage to Longview, Captain Lewin was struck by how close this typically underpowered, middle-aged cargo ship had come to making headlines around the world. There was only one small line separating the anonymity of the *Tai Shan Hai* as she plied the planet's oceans from an environmentally disastrous oil spill generating 24/7 TV coverage and costing millions of dollars. And that was the expertise of that small group of sea captains who comprise the Columbia River Bar Pilots and the pilotage system created by them at the mouth of the Columbia.

This human expertise and technical capability embedded in the pilot transfer system made three critical accident-preventing contributions to the *Tai Shan Hai* incident. First, there was Captain Chuck Lane instructing the ship to stay offshore and use her engine to take some strain off the anchors. And then came the extraordinary flying capability of *Seahawk* pilot John Glen delivering bar pilot Mike Dillon in a 70-knot wind to the deck of the *Tai Shan Hai* just before a huge swell washed over her decks. Lastly, Captain Dillon's experience in worse North Pacific storms enabled the ship to ride out the storm and eventually move out of the shadow of Peacock Spit and safely offshore to deep water.

In a thorough post-incident investigation, Washington's Department of Ecology Spill Prevention Section commended the bar pilots on all three counts. The report also admonished the *Tai Shan Hai's* captain for poor communications procedures with pilots and English-language capability, failing to more fully ballast the ship "to improve its handling capabilities in anticipation of storm conditions that were forecast prior to his ship departing San Francisco," and inadequate planning for a bar crossing requiring special considerations.

CHAPTER 7

CAPTAIN DILLON'S NEAR MISADVENTURE

JUST BEFORE MIDNIGHT on February 12, 1994, Captain Mike Dillon was descending the pilot ladder when it was bunched up the side of the *Hyundai 106* by the pilot boat as it rode a huge swell and collided violently with the Korean car ship that Dillon had just piloted outbound across the bar to sea. As the pilot boat fell with the swell, it whipped out the ladder's slack and shot Dillon into the Pacific Ocean like a spit wad at the end of a rubber band.

Trying desperately to hold onto the manrope, Dillon had been launched upside down and straight up, reaching the top of his airborne climb eye level with the shocked Korean crewman on the main deck who had just been assisting with his disembarkation. Captain Dillon then made the long dive into the Pacific just forward of the pilot boat *Columbia*. He'd been extremely lucky not to land headfirst on the deck of the pilot boat, which had moved slightly aft following the collision. Now nearly 20 feet under water, Dillon had to worry about avoiding the blades on the pilot boat's two propellers. Holding his breath and looking up, Captain Dillon could see the *Columbia's* silhouette backlit by the ship's spotlights as the pilot boat passed over him. At the time, he had no idea that her engines were jammed full ahead and that she would be unable to effectively conduct a search.

Having avoided the pilot boat's twin props, Dillon now had the ship's propeller – "the big boy" as he recalled – to worry about. By now, Dillon had hit the side of the ship twice and could feel the concussions of the propeller. In an instant, he knew that the *Hyundai 106* had failed to implement standard man overboard procedure. "If you lose someone overboard, we're trained to do four things immediately: stop the shaft, throw the rudder hard over so the prop wash keeps him away from the propeller, assign a lookout and throw life rings," said Dillon. "But he didn't do any of that. He just kept on going."

Knowing he might be only seconds away from being pulled into the eddy circulating around the 16-foot bronze propeller and certain death, Captain Dillon readied himself for what he hoped would be a last chance opportunity to push off the ship's side. "I felt the side of the ship ripping by me, and with my right leg I just kicked and I swam like hell, frog-kicking as fast as I could," remembered Dillon. "It felt like a Walt Disney ride, but I got far enough away."

Thrashing out his best crawl stroke fully clothed and slowed by his float coat, Dillon could feel the "shoo-shoo-shoo-shoo-shoo" cavitational concussions of the prop and momentarily feared the worst. But he had missed the bronze "big boy" and was now being thrust away from the vessel by the force of the prop wash. By the time he got himself turned around and reached the top of the swell, he could see that he was now a full ship's length away. He also saw the *Columbia* still alongside and wondered why she wasn't yet away from the ship searching for him. Only later did he find out that the *Columbia's* engine controls had been damaged in the collision and her engines had been jammed full ahead.

As he hit the top of another crest, Dillon could see that the *Columbia* was now headed his way and he began to think about being seen in the darkness. First he checked the strobe light connected to the shoulder of his float coat, which was supposed to deploy automatically upon contact with seawater, but "the damned thing crapped out." He then checked his handheld VHF radio, but it wasn't submersible and died in the ocean. He threw both away in frustration. And to top it off, Dillon had lost most of the reflective tape on the front of his float coat when he was raked down the side of the ship.

The next time Dillon rose to the top of a crest, he could see the *Columbia* turning away. He began to shout, but soon figured it was no use in the noise of the thundering clash of wind and sea. So he reached into his emergency pouch for two six-inch plastic light sticks which he bent to break the inner vial releasing a chemical that activated a florescent green glow. Dillon frantically waved his arms, a glow stick in each hand, and splashed the sea, hoping that the reflection of green glowing through the seawater would attract attention. He saved his energy for the most vigorous efforts at the top of each crest, knowing that he was invisible in the trough between 30-foot swells. Atop each crest, he looked anxiously as he waved toward the *Columbia,* but could see no sign that the crew had spotted him. By the third crest, she was headed away from him and by the fourth, she was over the horizon. "That's when I realized that they weren't going to find me. I was out there for the long haul," said Dillon.

He knew the big threat now was hypothermia and he turned his attention to his equipment. His Mustang Ocean Class Flotation coat with its 15 pounds of positive flotation was working well. Dillon was floating comfortably and vertically. The Mustang was equipped with a beavertail, a neoprene flap that extends down from the back and is designed to come up between the legs and fasten in front, preventing heat loss. Dillon fastened the beavertail, tightened the hood on his fleece parka, and pulled the Mustang's wrist straps tight. Knowing he needed to maximize any positive insulation factor, he kept his leather gloves and shoes on.

With the sea temperature at 46 degrees, the insulation from his float coat and other gear was critical to extending his life expectancy beyond mere minutes. Optimistic by nature, Dillon knew it was unlikely that the Coast Guard would attempt a helicopter rescue in 30-foot seas and 60-knot winds, but he soldiered on, telling himself jokes and trying to conserve energy. He also guessed – correctly as it turned out – that the Coast Guard's heaviest weather motor lifeboat would launch, but would not be able to cross the bar because of the treacherous seas breaking across it.

Still, as Dillon drifted, he clung stubbornly to the hope of some rescue. He prayed, thought of his wife Karen and their two kids, and even of readying himself to grab onto Buoy No. 1 if he drifted three nautical miles north of where he had been launched into the sea by the pilot ladder. From his training, Dillon knew that he stood a better chance of surviving clinging to the buoy and that an outbound ship would likely see him the next morning.

Despite his best efforts and a positive attitude in the darkness and surging seas, twice he felt the gathering strength of the cold water shudder through him. "It was like an upwelling of terror, like a cold shuddery feeling, and I knew that it was the kind of thing that if I let it overtake me it would be kind of an uncontrollable devastating type of fear. I never felt that before in my life. And I knew it was something I didn't want to have to deal with. It did affect my stomach – your stomach gets really nervous. I felt like it was going to start in my feet and come and overtake my body. A dark, cold, helpless type of feeling."

Each time the cold fear seemed to take hold, he tried to shake it off and tried to tell himself another joke and concentrate on feeling warm.

When he felt the need to relieve himself, he held off and waited for the next cold shudder, knowing that urinating between his coat's beavertail flap and his skin would have much the same warming effect as peeing in a wetsuit. "I learned that during scuba training in cold weather, and boy, I saved it up. It was great – you're damned right! I'll tell you what, when you're saving your life, you'll do anything you can to make it work."

By now, Captain Dillon had drifted 1.2 miles north and had been in the water 40 minutes. He should have been unconscious, but his level of fitness, concentration and positive outlook had saved him so far. Ashore, a life flight on alert for his rescue was cancelled because there was now no chance of survival. As he topped another crest, Dillon saw a dull light in the distance. Knowing that it wasn't the moon, he figured it had to be a helicopter.

Sure enough, Dillon soon saw the spotlight from the Coast Guard's Dolphin HH65A helicopter. Dillon grabbed his last light stick and tried to bite off the foil wrapper with his teeth chattering. Laughing once again, he finally managed to pull off the wrapper, snap the stick to activate the light, and wave it in one hand while he waved his still functional flashlight in the other.

As Dillon grabbed for the flashlight, he noticed that his hand seemed to close around the handle as if in slow motion. He knew in that moment that hypothermia was slowing him down, but he felt adrenaline come to his rescue as he waved his lights, sometimes shining one at the one piece of reflective tape still on his chest. Trying again for some attention-grabbing reflection as the helicopter flew over, he thrashed the sea with his arms, exhorting himself to hold on tight to the two lights. The helicopter came close, but seemed to go by so fast that he thought, "Well, that was my big chance. Son of a gun."

Aboard the helicopter, however, the winchman had seen a flash of light. Circling back a second time, the crew saw another flash which was exceedingly rare in those conditions. According to Lieutenant Mark Jessen, who was in command trying to circle in strong winds, "You get blown all over the place. I started this turn to go back around and try to find the target again. That was just luck. It just doesn't happen."

Several days earlier in a training exercise, Jessen's crew had tossed a training target into heavy seas in broad daylight and, despite multiple attempts, never found it again. Jessen knew this rescue mission was a longshot in stormy, dark conditions. He considered the chances "pretty much nil at night, in a rain storm, and 40-foot seas. And we found him three times. God was shining on that man. There was no reason we should have found him."

On his second pass, the crew dropped a smoke flare close to Dillon and then began a third circle while the crew prepared to execute the rescue using a rescue swimmer and the winch. But the wind pushed the helicopter nearly two miles downwind. Jessen then had to increase his speed from 70 to 90 knots to make any headway. Once back above Dillon, rescue swimmer Steve Wanlass was lowered into the ocean clad in wetsuit, facemask, and fins. As Wanlass approached, Dillon wanted him to know that his target was calm not crazed, so he quipped, "What are you doing out here on a night like this?" Pleasantries were quickly exchanged and Wanlass set about getting Dillon into a harness and hooking them both to the wire dangling from the helo.

Meanwhile, with his rescue swimmer still "on the hook," Lieutenant Jessen fought to keep the helicopter somewhere close to steady. To keep the lights from being diffused by all the sea spray, he had to stay low and fly up and down the 30- to 40-foot swells. "I had to climb as the seas came. It wasn't as choppy as you'd expect, because the wind was so strong, it kind of flattens things out. But the swells were magnificent. It was like flying over rolling hills."

After Dillon and Wanlass were hooked in, but before the signal to winch them upward, another swell caught them and pushed them to what Dillon recalled was within three feet of the helicopter undercarriage. "We both ducked under," said Dillon. "Our first thought was that the helo had lost power. We thought we were going to get hit." While Jessen pulled up to avoid the peak of the swell, the hoist operator spooled out wire quickly to prevent the pair from being jerked out of the ocean while descending into the trough.

Then the hoist began. Once out of the sea and into the biting wind, Captain Dillon felt both cold and numb. In fact, he was so numb as he reached the helicopter door that he couldn't lift his knees up to enter, so Wanlass just shoved him in. As the helicopter lifted up and away, all aboard whooped for joy. Said Dillon, "I mean, I was happy. They were happy."

Dillon was flown to the Astoria Airport and whisked by ambulance to the hospital. Upon arrival, his core body temperature was less than 90 degrees. It took 14 hours to slowly warm him back to the normal 98.6 degrees. He spent another four days at home under doctor's orders waiting for his heartbeat to return to normal.

When he returned to active piloting, Captain Dillon brought with him a new focus on safety. He spent $1,200 on one of the best float coats on the planet, an English-made SeaSafe with 30 pounds of water-activated flotation, plenty of reflective tape, a water-activated strobe light and a radio beacon. He also upgraded his other gear, adding larger light sticks and a waterproof radio. Within the bar pilot organization, Dillon threw himself into the task of figuring out how to replace the station boats *Peacock* and *Columbia,* the only pilot boats he had known since joining the bar pilots in 1989. He was one of a trio of bar pilots who tested helicopter operations in France in 1996, which occurred before he went overboard a second time in disembarking an outbound ship in heavy seas. This time, his new float coat inflated instantly and the pilot boat successfully recovered him.

For several years after his brush with near death, Captain Dillon spoke widely to the press and to fellow pilots about the importance of safety and good equipment. Along with Captains Roger Nelson and Gary Lewin, he helped convince a majority of the bar pilots to support the change from the station boat system to the combination of a dedicated helicopter and highly maneuverable jet-powered aluminum pilot boats. And after the hard fought rate proceeding in 1998-99, Captain Dillon was proud to cast one of the votes in favor of the new helo/fast boat system as one of nine appointed members of the Oregon Board of Maritime Pilots.

Now retired from his search and rescue career in the U.S. Coast Guard, Mark Jessen remembers the Columbia River Bar as "the most dangerous in the United States and Astoria has some of the toughest flying."

The immediate success of helicopter operations was gratifying to Captain Dillon and its capability in difficult weather eventually won over that minority of bar pilots who either doubted or opposed its introduction. Dillon used the helicopter whenever possible, but continued to be haunted by the pilot ladder incidents that had nearly claimed his life. As he puts it, "The commute is hell on the Columbia River Bar. The highest danger is just getting to and from the ship."

Looking elsewhere for the opportunity to pursue his profession in a less dangerous environment, Dillon secured a position with the Galveston-Texas City Pilots Association in Texas in 2004. He plans to retire in 2017, but will always remember the Columbia River Bar as "the damnedest, toughest place to pilot in all the world, where you've got to be right at the top of your game."

CHAPTER 8

CAPTAIN QUINN'S PULLING BOAT LEGACY

IN 1951, WHEN HE JOINED THE BAR PILOTS, Captain Edgar A. Quinn epitomized the ideal recruit for the world's most dangerous pilotage ground. He was 40 and had over 20 years of experience as a seagoing mariner, including 11 as a master with Matson Navigation Co. His work as a ship captain for Matson was interrupted by World War II, when Captain Quinn commanded troop ships in every theater except the Mediterranean.

Once a bar pilot, Captain Quinn served as president of the organization for five years, as secretary for three years, and as pilot boat manager. He was also active in local civil affairs, serving as a member and later chairman of the Clatsop County Library Board, as a founder and director of the Columbia River Maritime Museum, and was active in his church and other organizations.

An excellent athlete, Captain Quinn excelled in rugby, soccer, and cricket during his London boyhood. Like most bar pilots, he maintained a high level of fitness, which provided an important margin of safety for boarding and disembarking ships in rough conditions.

Captain Quinn's fitness and skills as a bar pilot were put to a well-publicized test in the fall of 1962. On September 28, rain squalls blanketed the face of the Pacific just before midnight when Captain Quinn, then 51, descended the pilot ladder lashed to the steel hull of the Japanese freighter *Olympia Maru* and jumped into the waiting 16-foot pulling boat. Crewing the pulling boat that day were seamen Don Nelson, 34, and William Wells, 35.

By this time, the traditional pulling boat was fiberglass, made from a mold of the traditional wooden hull, and fitted with a 10-horsepower outboard motor. But two pair of oars remained standard equipment, lashed just inboard of the two rowing stations.

Quinn took his seat at the stern of the small craft, wiping from his face the cold spray that whipped across the ocean as a 60-knot gale began to sweep ashore. He was looking forward to a hot cup of tea aboard the *Peacock,* which was just a few hundred yards away with her bow facing into the oncoming storm.

A sudden freak wave caught the pulling boat, juggled it up to the top of its crest and then turned the small craft "turtle" or upside down and slammed her into the trough. "We righted the boat finally by hauling on the keel line, which had been rigged for just such a disaster," Quinn recalled later. The pulling boats used by the bar pilots were designed to be righted after capsizing, but it was no easy task at sea. On this stormy night, Quinn, Nelson, and Wells swam to the same side of the pulling boat, positioned themselves from bow to stern and then grabbed the line laced through the keel and jerked the boat upright.

Once aboard, the sheet metal bailing can that could double as a sea anchor was put to work and Captain Quinn surveyed the scene. "We were awash and I could not see the *Maru* or the *Peacock* and I knew then that we were on our own," he recalled.

Just like what happened to the *Iowa* more than two decades earlier (see Chapter 9), the gale-force winds out of the southwest and the strong northerly set of the current drove the pulling boat north all through the night. The capsizing fouled the outboard motor so Don Nelson rowed for three or four hours to keep the boat running into the wind and sea. Bill Wells had started the day with a touch of the flu that worsened in the terrible conditions and rendered him too weak to help with the rowing. Captain Quinn shouted encouragement often to seaman Wells, who seemed to be growing weaker.

As Nelson began to tire from rowing, he was able to rig the bailer as a sea anchor that kept the boat headed into the seas. But when the handle broke and the sea anchor disappeared, Nelson went back to his oars until daybreak. As morning light emerged, the three could make out the Washington shore through the driving spray, but the huge storm-driven swells made the idea of heading ashore suicidal.

Bill Wells grew weaker throughout the night and into the morning. At about 10 o'clock, he crawled forward to the bow where he shivered in the swirling seawater. Wells died at approximately 10:30 a.m. Using the piece of rope that had been tied to the sea anchor, Nelson lashed Wells' body tightly to the transom while the storm continued to drive the boat north for another 18 hours under rainy gray skies.

With no food or water, Quinn and Nelson began to suffer from exposure and became ravenously thirsty. "We stuck out our tongues to catch rainwater and we kept telling each other we were going to make it," said Quinn.

As another September night began to fall upon them, the swells had moderated and the pair headed toward the lines of breakers along the shore, rowing frantically. Once in the breaking waves, which extended over several hundred yards, the pulling boat rolled over multiple times and had to be repeatedly righted and reboarded. Finally, after yet another capsizing, their feet touched bottom while trying to re-right the craft and they started to wade ashore. Incredibly, the pair had spent more than two hours working their way through the breakers, riding and reboarding their capsized craft over a dozen times. But their journey was not over.

Quinn and Nelson were so weak that neither could make it on his own and they had to wrap their arms around each other to make that last distance through the surf zone. Once ashore, they supported each other through a three-mile hike to a motel which was emitting a bit of light. It was tough going. Nelson had lost his boots in the ordeal at sea and nearly broke his ankle in one of the collisions with the pulling boat. It was nearly midnight when Quinn and Nelson knocked on the door of the motel. "A woman opened the door with eyes as big as half dollars looking at us beach rats," said Nelson. She had heard about their plight on the radio, brought them inside, sat them down by her fire and covered them with blankets. She brewed a pot of coffee, but neither man was able to hold up a coffee cup and so she spoon-fed it to them.

During one of the multiple capsizings while coming ashore, William Wells' body had broken loose from its lashing and had drifted away. Several days later, the body washed ashore on a beach north of the spot where the pulling boat was later recovered.

A massive search for the three missing men had been launched from Astoria, and the *Olympia Maru* had turned back from the start of her return voyage to Japan to join the freighter *Michigan* in the search. These vessels teamed up with the pilot boats *Peacock* and *Columbia* and the Coast Guard Cutter *Yocona* in a desperate search for the three.

Quinn and Nelson had come ashore near Grayland, Washington, some 25 miles north of where the pulling boat had originally capsized. The woman who was tending to them at the motel called the local sheriff, and a deputy soon arrived to deliver the two men to St. Joseph's Hospital in Aberdeen. Sitting in the back seat, Don Nelson became concerned for his safety as the police cruiser reached speeds in excess of 100 miles per hour. Nelson tapped the deputy on the shoulder and said, "Sir, I don't mean to tell you your business, but I didn't survive this boat wreck to be killed by a car wreck." The deputy sheriff slowed down and Nelson sighed to himself in relief, "We're not dying."

Quinn and Nelson spent two days in Aberdeen and then were transferred to St. Mary's Hospital in Astoria. While there, one of Don Nelson's friends snuck two quarts of Canadian Club into his room. When an elderly nun discovered his stash, Nelson thought it was sure to be confiscated. Instead, she smiled broadly and said with a wink, "This is fine medicine if it is used right."

Incredibly, Nelson was back on the bar within a week after his discharge from the hospital. Captain Quinn, however, who had been wearing the then-standard pilot uniform whipcord trousers and matching jacket, lost most of the skin on his upper legs down to his knees from the combination of the abrasive fabric and salt water. He was off for several months as he grew back that skin.

Bill Wells' wife was pregnant with twins when he died. Don Nelson had nightmares about the loss of his shipmate and his own close call with death for several years.

CHAPTER 9

THE *IOWA* DISASTER

JANUARY 11, 1936, WAS A FAIRLY ORDINARY winter Saturday in Longview, Washington, on the lower Columbia. The steamship *Iowa* was moored at the Weyerhaeuser mill dock loading a Pacific Northwest product mix: canned salmon and fruit, flour, matches, shingles, and lumber. The two hatches aft were fully stowed, but the cargo space in the two forward hatches was reserved for additional cargo to be taken on in California ports. The steamer was 410 feet in length, 5,724 gross tons, and carried a primarily Portland-based crew of 34. A deck cargo of lumber both fore and aft was loaded last and the *Iowa* departed Longview early that evening.

The combination of the empty cargo space forward and the heavy timbers on the after-deck resulted in a nine-foot drag at the stern: the *Iowa's* draft was 26 feet at the stern but only 17 feet at the bow. It was not unusual, however, for coastwise vessels (U.S. flag ships trading along the U.S. Coast) to transit between ports with substantial stern drag. Ultimately, the decision whether to restow to adjust the ship's trim was left to the sound judgment of her captain.

The *Iowa's* master was a man with considerable seagoing experience. Captain Edgar L. Yates, 68, was a native of England who earned his master's license in his native country before immigrating to the U.S. in 1916 to accept a position as a captain with Oriental Navigation Company, running vessels in the trade between New York and the Far East. After a dozen years operating out of New York, Captain Yates signed on with States Steamship Co. in 1928. One of his first assignments was as captain of the *Oregon,* and he eventually moved to Portland, where he lived with his wife Mary on southeast 21st Street. Built in San Francisco in 1920, the *Iowa* was originally christened the *West Cadron* and engaged in the trans-Pacific trade from her home port of Portland as part of the fleet operated by Columbia Pacific Shipping Co. When States Steamship Co. acquired that company in 1928, the steamer's name was changed to *Iowa* and she continued in the trans-Pacific service several more years. In 1934, the *Iowa* suffered major damage in a collision with the *Heiyei Maru No. 12* off Hakodate, Japan. Following her repairs, the *Iowa* returned to service in the U.S. intercoastal trade and was scheduled to discharge her West Coast cargo in New York in a few weeks' time.

With her deck load of lumber secure late Saturday, river pilot Captain Stewart Winslow boarded the *Iowa* while she lay alongside Weyerhaeuser mill dock and then supervised her downriver transit to the mouth. While enroute, Captain Winslow asked Captain Yates if he wished to take on a bar pilot. "No," he replied, exercising his option as the captain of a U.S. coastwise vessel who held both a U.S. master's license and a pilot's license for the Columbia River. Based on the weather at Astoria, conditions for the bar crossing appeared favorable, but storm warnings had been posted for 36 hours. The wind was easterly and the tide was flooding in with high tide at 2:50 a.m. It was rainy, but the range lights were visible when Captain Winslow debarked the ship off 14th Street, climbing down the pilot ladder and boarding a launch for his return ashore.

The *Iowa* left the lights of Astoria and proceeded nine miles downstream to the vicinity of Buoy 12, just inside the bar and one mile from the Cape Disappointment Station where Coast Guardsman Craig was on lookout from midnight to 4:00 a.m. The weather was overcast with rain and hail, but most ominously, the warned of storm was now a reality. A gale was blowing in from the southeast with winds increasing from 9 to 11 Beaufort scale, which was a range of 47 to 72 miles per hour. The seas were breaking on the bar, but Craig watched as the *Iowa* proceeded on to meet the incoming gale and elected not to abort her bar crossing and anchor in the relative shelter of Baker Bay. Rather, she fought the wind and waves and slowly proceeded to Buoy 10 where she seemed to pick up some speed as she headed out toward Buoy 6, appearing to be on her regular course.

Approaching Buoy 6, where the channel makes a dogleg turn to the southwest, Craig observed the *Iowa* appear to hesitate, although he wasn't sure whether she actually did or seemed to because she was sailing directly away from him. To Craig, the *Iowa* appeared to be bucking the wind and current. Rain squalls periodically obscured his view, but at 2:45 a.m. he noted that the *Iowa* seemed to be drifting northward. Craig last saw the outbound vessel just before 3:30 a.m. when "a hell of a squall came by" and he lost sight of her. At the time, he could see her stern light as she made the turn at Buoy 6 and, despite her northward drift, he thought she would clear the sands of Peacock Spit and make it to the safety of deep water.

At 3:12 a.m., 18 minutes before Craig's last sighting, *Iowa* radioman Frankie Caldwell, a Portland native and talented amateur boxer, was at his post. Like all aboard, he was unaware of any impending trouble. At 3:12 a.m., he radioed the States Steamship Co. office in Portland, "I have nothing for you." Everything seemed normal until 23 minutes later when the *Iowa* collided violently with the hard sand of Peacock Spit. In a short space of time and unbeknownst to the crew, the ship had been driven a distance of more than two miles north of the navigation channel.

The fateful call came at 3:45 a.m. Radioman Caldwell urgently tapped out the Morse code message: "SOS. On Peacock Spit, breaking up." Wind, wave, and a strong northern set or current were the causes of this fateful deviation off course. And the crew had apparently been caught unawares. No flares, rockets, or other distress signals were given. The ship's anchors, which if deployed with her engines running might have arrested the drift ashore, remained in their hawse pipes. At 3:55 a.m., there was one last radio transmission from Frankie Caldwell: "SOS."

Hearing the distress calls, the crew at the North Head lighthouse, some three miles north of the stranded ship, sprang into action. After relaying the first radio message about the stricken ship, the lighthouse crew trained its telescope toward Peacock Spit. The *Iowa* was first sighted about 4:30 a.m. by Berry Bray, assistant keeper of the lighthouse. She was drifting helplessly and he immediately notified lighthouse chief A. G. Siniluoto. The pair were soon joined by junior meteorologist Nino Sunseri, his wife, and his assistant Charles Hubbard in the lighthouse tower. In silence, they watched the *Iowa* slowly stop her drift. The eerie silence was broken by a call from the Naval radio station at Astoria. The wire advised that the *Iowa* had issued an SOS, but all attempts to contact the steamer had been futile. The lighthouse was asked to communicate with the *Iowa* by means of signal light as the vessel's radio was obviously out of commission.

Using the storm warning lights on the lighthouse tower, Sunseri blinked out a Morse code message: "Ans. Ans. Ans." meaning "Answer." Straining their eyes as they peered into the rain and fog, the lighthouse crew "saw a feeble light from the top of the pilothouse, barely discernible through the fog-enshrouded air," described Sunseri. "Then a heavy rain squall came up and all vision was completely obliterated."

As dawn was approaching, the lighthouse crew ran up an international code flag. "Those poor devils on the boat must have seen our flag," Sunseri related, "for they immediately sent up some of their own." The lighthouse crew could see the *Iowa's* pennants flapping in the fierce wind, but could not make out what flags they were or the intended message. While assistant meteorologist Hubbard had his eye glued to a telescope trained on the *Iowa,* he saw a crewman emerge from the pilothouse and head for a ladder to the foremast. He had taken only a few steps when a huge breaker engulfed the entire ship and tossed the man like one of the matchsticks in the *Iowa's* cargo holds into the raging sea.

The rain squall had now subsided, affording the five in the lighthouse tower a clear view of the *Iowa* and her condition. The ship's entire deck was bare: no lifeboats or deck load of lumber remained. The lifeboats could be seen bouncing on the wild seas without any sign of life aboard. Huge incoming breakers pounded the *Iowa* one after another. Within a matter of minutes, the stack and pilothouse were swept away and the water line ascended rapidly to the vessel's upper decks. The pounding by tremendous waves continued unabated until the *Iowa* heaved violently, appeared to float upward momentarily, and then slipped completely out of sight, ex-

THE *IOWA* DISASTER

Within hours of the *Iowa's* sinking, only her mast, the top of the wheelhouse and two stacks remained visible above the seas that had broken her in two and killed all 34 men aboard. *Columbia River Maritime Museum photo.*

cept for the mast, which kept a silent vigil over the spot where most of the 34-member crew was entombed.

The Coast Guard Cutter *Onandaga* received a radio call that a ship had wrecked at the mouth of the Columbia River at 4:18 a.m. and that the winds were then blowing 45 miles an hour. After trying unsuccessfully to radio the *Iowa,* the *Onandaga* prepared for sea, cleared the dock at 6:24 a.m., and headed downstream with an ebb tide.

Lieutenant Commander R. S. Patch, the officer in charge, described the ordeal of attempting to cross the bar in these conditions:

> *The wind and current forced us out of the channel and we fought our way across the bar. The engines were turning 75 revolutions, which should produce a speed of 11.3 knots. It took us, however, 45 minutes to make 1500 yards.*
>
> *My men had to secure the boats five or six times as big seas boarded us. Visibility was not more than a mile and we made for the lightship but saw no one. I figured the ship if it struck on the south side, had carried north so I changed our course and going north I could see, at 9:10 a.m., a mast smothered in breakers. Closer, I could see the stack, bridge, foremast and windlass on the forecastle head. In the trough of the seas I could get a glimpse of the tops of the ventilators and figured the decks were 10 feet under water then.*
>
> *We estimated the seas to be 60 to 75 feet high and my fathometer at times showed a variation of from three to 15 fathoms between the crests and troughs. At 11 a.m. a particularly black squall blotted out the ship and we could see nothing. We first thought we saw three men on the foremast but later discovered they were not there and what we saw were rigging blocks.*

The *Onandaga* suffered considerable damage in her unsuccessful attempt to reach the *Iowa*. Two of her lifeboats were badly damaged and the steel mounting of an anti-aircraft gun was sheared away.

Despite the loss of communications from the storm, word of the loss of the *Iowa* spread quickly. For a while, there were rumors of three other ships lost, but each was accounted for within 24 hours. At the same time, the deadly Pacific began to release some of those who had perished on that early Sunday morning. Six badly battered bodies, each having suffered multiple fractures in the pounding seas, washed ashore with the incoming tide on Monday, January 13.

The same high tide carried ashore portions of the ship's cargo. The beaches north of the river's mouth were soon swarming with beachcombers. One man managed to supplement the family larder with over 20 cases of canned salmon. Dozens of others salvaged the still pristine center portions of sacks of flour. Only the outer few inches were penetrated by seawater and the contents further inside were untouched. The ship's two million board feet of lumber cargo presented the largest salvage opportunities. Many timbers were splintered like matchsticks in the pounding surf, but much came ashore intact or at worst oil stained. A local newspaper reported that enough lumber had been salvaged to start three new lumber companies.

While the salvagers stripped the beach bare of the *Iowa's* remnants, radioman Caldwell's mother hurried to Peacock Spit where she watched for any sign of her son and his ship. For two days, she walked back and forth through the cold wind and drenching rain, refusing offers of food and shelter through teary eyes looking out to the sea that held her son. His resting place was marked only by the gray mast that served as a lonely sentinel to the tragedy.

Ultimately, the Pacific yielded only 10 of the 34 members of the *Iowa's* crew. And this despite the fact that the vessel split in two and the stern was discovered 125 yards from the bow, still marked by the broken mast and visible at low tide. The jetties were still causing substantial accretion of sand on Peacock Spit and Washington's north shore. Much of that accretion remains, making it probable that there are still the remains of 24 of the *Iowa's* crew entombed in the sand of Peacock Spit.

The immensity of the disaster stunned the nation. How could a modern U.S. flagged ship, operated by one of the nation's steamship lines, sink at the mouth of an American river with all hands lost? Five maritime unions that had members aboard the *Iowa* called immediately for a full investigation and charged that concern over profits rather than safety had contributed to the decision by Captain Yates to attempt a bar crossing in the face of storm warnings without a bar pilot.

Labor leader C. O. Hunter charged, "The belief is increasing that the loss of 34 lives was due to first consideration being given to profits. The boat was attempt-

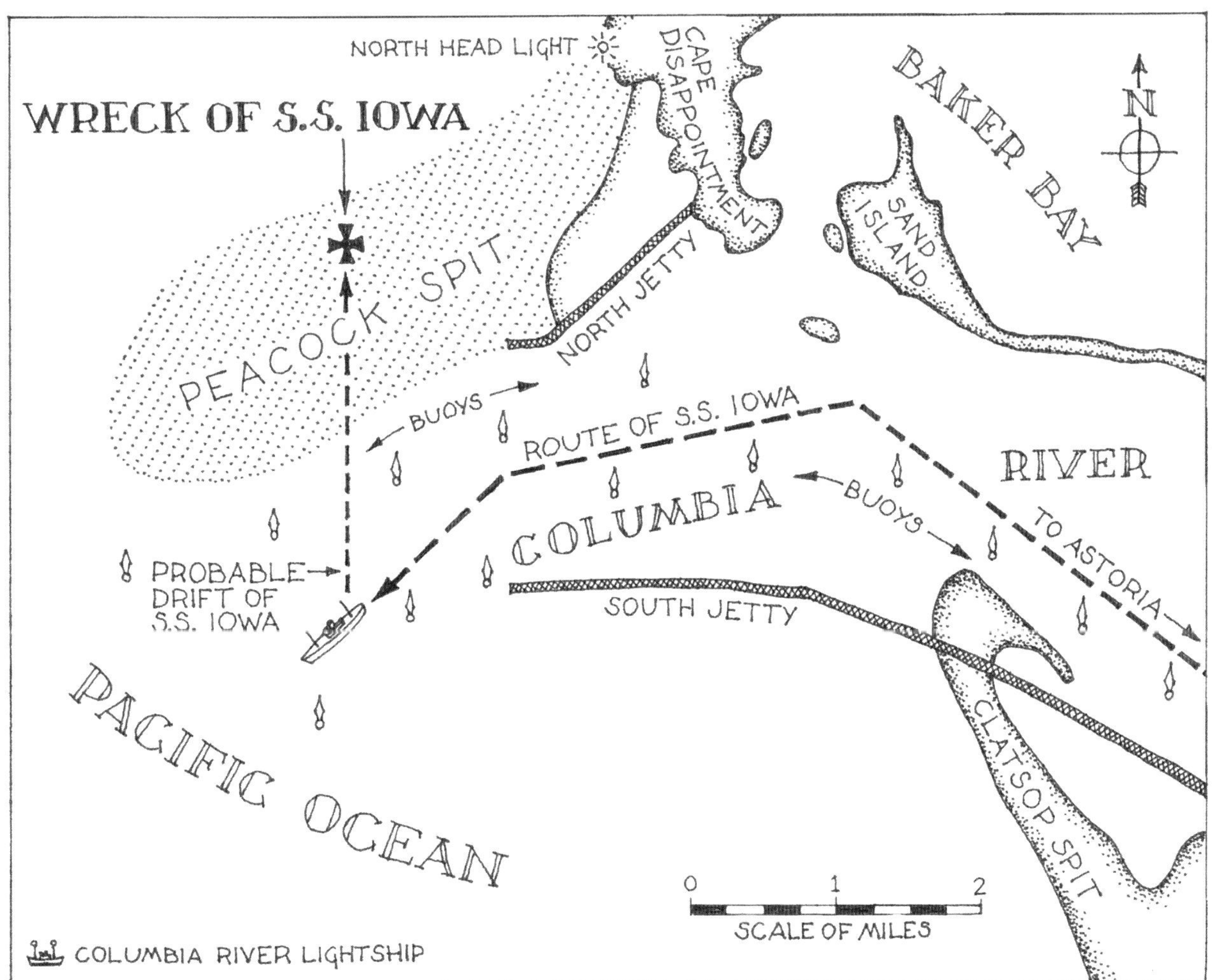

WHERE FREIGHTER *IOWA* LIES IN A SHALLOW LOCKER

This recreation of a hand drawn illustration in the *Oregonian* shows the location of the *Iowa's* grounding, some two miles north of where she intended to be in the middle of the navigation channel. This illustration ran under the headline "Where Freighter Iowa Lies in a Shallow Locker."

ing to pass out of the bar without a bar pilot. The ship encountered an unusual storm, and it is possible that no pilot would have been able to keep her off the spit, but it is probable that a competent bar pilot would have declined to take the ship out."

"When it left Astoria dock about midnight the storm was raging, storm signals had been up for 36 hours," continued Hunter. "With a crew aboard, delay of a day of course meant considerable expense. It is a common practice to risk the safety of men in order to save money."

Pilotage rates for the Columbia River Bar were $1 per foot of draft plus 1¢ per net ton. For the *Iowa,* with a draft of 26 feet and at 3,564 net tonnage, the fee for a bar pilot would have totaled $61.64.

Two days later, the *Oregon Journal* called for an official inquiry into the wreck of the *Iowa,* noting that the central question to be examined was the validity of the charge by organized labor that, if the *Iowa* had taken a bar pilot, would the catastrophe have been averted? The editors declared, "If this point is made, there can only be one result: to require pilot services for ships under the American flag, as is now the requirement for foreign vessels."

Two weeks later, the director of the U.S. Bureau of Navigation, J. B. Weaver, convened a hearing in the U.S. District Courthouse in Portland. Three bar pilots were called to testify. Captain William Hall was called first. He was the bar pilot on the *Koyei Maru,* which was the last ship to cross the bar before the *Iowa's* fateful voyage. Captain Hall testified that when the *Koyei Maru* crossed the bar at 7:30 p.m., some four and one-half hours before the *Iowa* departed Astoria, the wind was blowing 40 to 45 miles an hour out of the south. In answer to questions from Director Weaver, Captain Hall testified that he would not have taken the *Iowa* out with her power and size later that night.

In his testimony, Captain Hall also expressed qualified support for a proposal to establish a lookout station for reporting bar conditions. The endorsement of an observation point above Peacock Spit was limited because it would be difficult to truly ascertain bar conditions – specifically wind speed and wave height – from a lookout hundreds of feet above the surface of the river. The best place and time for decision is when approaching the mouth of the river near Buoy 12 where the forces of wind and wave can be felt and there is safe anchorage available if a bar crossing is inadvisable.

The other two testifying bar pilots, Captains Robert Marquardt and Oliver P. Rankin, both testified that they would not have attempted a bar crossing with the *Iowa* in the face of the incoming storm. Captain Rankin described the strong northerly set or current that develops at the mouth of the Columbia after a gale has blown out of the south and that it is not advisable to attempt a bar crossing against the combination of a strong northerly set and stiff southerly winds. He also expressed concern about the *Iowa's* power in rough conditions, noting that when he piloted the vessel outbound on Christmas Eve he had advised the master to get more steam.

The total loss of the *Iowa* imposed immeasurable pain on 34 families, the wives, children, parents, and siblings of the 34 seamen who perished. That pain was only sharpened on the two-month anniversary of the disaster when the *Iowa's* owner, States Steamship Co., seeking financial refuge in a federal statute first enacted in 1851, filed a petition in federal court seeking to limit its liability to the value of the wreck and her pending freight, a sum of $10,624.76. The statute allowed a ship owner to limit its liability to the post-casualty value of the vessel plus freight provided that neither shipowner negligence nor vessel unseaworthiness was the cause of the loss. More than 30 cargo claimants and the estates of seven deceased seamen, including the estate of Captain Yates, challenged the petition alleging multiple grounds of owner fault and vessel unseaworthiness.

The federal judge in charge of the case appointed a commissioner, Robert F. Maguire, to conduct a factual hearing and to make findings of fact and recommendations regarding disposition of the case. The hearing was exhaustive, consuming 28 day and 13 night sessions and generating a transcript of 32 volumes and 5,160 pages. The commissioner's detailed report was issued in April 1936, more than two years after the casualty. Maguire found no evidence to support the claims that the *Iowa* was dangerously loaded with excessive stern drag, that her steering gear, engines, and compasses were out of order, or that the ship was underpowered. In rejecting these claims, the commissioner emphasized the successful prior seagoing history of the vessel and the common practice of allowing stern drag during a bar crossing.

The commissioner did find that States Steamship Co. had failed to furnish the vessel with a Notice to Mariners advising of the loss of two buoys and the light on a third buoy and further that the vessel lacked a required communication system between the bridge and engine room. Both rendered the *Iowa* unseaworthy, but the commissioner nonetheless found that neither failure had any causal connection to the loss.

On the most critical issue – whether the *Iowa* should have taken a bar pilot – the commissioner rejected both arguments directed at this charge. First, the ship owner had no obligation to order Captain Yates to take a bar pilot as that decision was left to the discretion of U.S. flag masters who had the credentials to pilot their own ships. Second, States had fulfilled its legal obligation by engaging a competent shipmaster in Captain Yates to take charge of the ship's navigation.

Ultimately, Commissioner Maguire placed all of the blame for the total loss of the *Iowa* and her crew on Captain Edgar Yates:

> *When even the most friendly witness finds himself unable to say that a good seaman would attempt the passage, when everyone familiar with the conditions as they then existed or have been shown to exist, states that it was unsafe to attempt it, the Commissioner is forced to the conclusion that Captain Yates' action of putting out over the bar at the time he did was not a mere error of judgment but was in fact negligence. He was under no compulsion by orders, suggestion or otherwise to cross out that night. As he approached the mouth of the river the force of the gale must have been apparent. That he would face breaking seas and a strong northerly set, with the wind on his port bow, was unquestionably known to him. He could have safely anchored and awaited moderation of the weather, but for some reason he concluded to take a chance and it was fatal to him, his crew and his ship.*

Captain Yates' estate and the other claimants objected to the commissioner's finding, but their objections were swept breezily aside by District Judge James Alger Fee on the basis of what he characterized as the commissioner's "able report in which the questions are exhaustively discussed."

Judge Fee also adopted the commissioner's finding that the *Iowa's* owners had met their legal obligation by providing a competent master who was in complete control of the vessel's navigation, noting that the "owners in Portland would have no right to dictate" Captain Yates' actions nor could his error be imputed to the owners. Curiously, the district judge seemed almost to praise Captain Yates' negligence:

> *The gallant shipmaster who took the chance, relying on his skill despite the storm, compels admiration. He misjudged and failed. He expiated fault by going down with his ship according to the tradition of the sea. In the light of all the circumstances, his recklessness was the sole cause of the disaster. The testimony of the single-witness to the behavior of the ship confirms this conclusion.*

Judge Fee's decision exonerated States Steamship Co. from any liability beyond $10,624.76. After expenses, the 37 claimants recovered little or nothing from this fund.

Ultimately, despite the unequivocal conclusion of an encyclopedic federal court investigation finding that Captain Yates had attempted a bar crossing in conditions that no bar pilot would have considered, there was no action taken to change the statute allowing U.S. coastwise vessels to cross the bar without a bar pilot. To this day, that option remains unavailable to all foreign flag vessels, but is regularly exercised by the captains of U.S. flag oil tankers owned and operated by America's major domestic oil companies.

CHAPTER 10

CAPTAIN GEORGE FLAVEL: "IF I LIVE, I WILL RETURN."

WITHOUT QUESTION, CAPTAIN GEORGE FLAVEL is the most famous of the Columbia River Bar Pilots. He built the first pilot organization, fought off competitors, and amassed a colossal fortune while establishing the tradition that only experienced ship captains were qualified to become bar pilots.

Born on November 17, 1823, in Northern Ireland, George Flavel was a 25 year-old sea captain when gold was discovered in California. In 1849, Flavel purchased the brig *John Petty,* stocked her with merchandise for sale to the miners, and recruited a crew for the voyage from Virginia around Cape Horn to the West Coast.

But upon arrival in San Francisco, Flavel discovered that many ships had arrived earlier, overstocking the market. So Flavel sailed the *John Petty* north to the Columbia River, across the bar for the first time, and then on to Portland where he sold his cargo for a tidy profit. Captain Flavel then returned to California where he tested his own luck at gold mining, but he was unable to find a profitable claim. By the fall of 1849, he was in command of the sidewheel steamer *Goliath,* operating a passenger service mainly for miners between San Francisco and Sacramento.

In 1850, Captain Flavel became pilot and mate on the *Goldhunter,* a 172-foot, 511-ton steamship running between San Francisco and Portland. Based on his experience transiting the Columbia River Bar aboard the *Goldhunter,* Captain Flavel received a branch license to pilot the treacherous bar on September 22, 1850 — the first such license granted by the then Territory of Oregon and its newly appointed Oregon Pilot Commission. Later that fall, the *Goldhunter* was sold and sent to Mexico, and Captain Flavel returned to his job commanding the sidewheeler *Goliath* between the Columbia River and San Francisco. When the vessel was tied up in Astoria, Captain Flavel boarded at the Astoria Hotel, owned and operated by Conrad Boelling. It was there that George Flavel first met the Boellings' beautiful youngest daughter, Mary Christina Boelling, who would become his wife four years later.

Obviously impressed with the opportunities in Astoria, the entrepreneurial Captain Flavel decided to make it his home. As it happened, organized pilotage service at the mouth of the Columbia was in its infancy, sparked by the California gold rush. The first pilot boat was the schooner *Mary Taylor,* which began her service on Christmas Day, 1849. Owned by bar pilot Captain Jackson Hustler, the *Mary Taylor* operated for two years with five pilots, but George Flavel heard many complaints from fellow shipmasters that Hustler's bar pilots and the *Mary Taylor* were not providing adequate service.

In 1852, Captain Flavel partnered with Captain Alfred Crosby to purchase the schooner *California,* a 64-foot vessel then in San Francisco that had been built four years earlier in Stonington, Connecticut. Flavel's plan was to sail the *California* to Astoria and put her into service as a pilot boat in competition with Jackson Hustler and the *Mary Taylor.* The *California* entered the pilotage business in 1852 under the command of Captain Flavel and a team of four pilots, including partner Alfred Crosby and Captains Charles Edwards, Peter Ferchen, and George Wood. The competition with Captain Hustler was fairly short-lived. Already in poor health, Captain Hustler sold the *Mary Taylor* within the year, and it was converted to an oyster barge on Willapa Bay.

1852 proved to be an especially auspicious year for Captain George Flavel. In addition to marking his first year in the bar pilotage business, 1852 generated the harrowing experience that would bring him the early fame on which he built a most remarkable career. No single event played a larger role in building Flavel's reputation as a mariner. He was ultimately described in *Lewis & Dryden's Marine History of the Pacific Northwest* (1895) as "a man who amassed a colossal fortune and made himself famous in marine circles throughout the Northwest."

On January 28, 1852, Captain Flavel piloted the steam-powered *General Warren* across the bar on her outbound voyage for San Francisco. She carried 31 passengers, a cargo of wheat, and 800 live hogs. After completing the pilotage assignment, Captain Flavel debarked to the pilot boat *California* and left the *General Warren* in the command of Captain George Thompson. As she headed south on her course, a heavy southwest breeze sprang up that quickly escalated into gale-force winds. By midnight, the *General Warren's* foretopmast had been blown off and the wooden hull had sprung a leak. Wheat shifting in the hold clogged the pumps, so Captain Thompson decided to return to Astoria for repairs.

The *General Warren* was off the mouth of the Columbia River by daylight, but the storm and heavy seas prevented her from being sighted by the pilot boat *California* until midafternoon on January 29. At this point, the wind had died down, but there was already three feet of water in the ship's hold and a heavy ebb tide was sending waves breaking across the bar. When the *California* came within hailing distance of the *General Warren,* Captain Thompson called out a request that Captain Flavel come aboard. Captain Flavel quickly made his way to the *General Warren* in a small boat. Once aboard, Captain Thompson took him to one side and said, "We have three feet of water in the hold. We won't live 'til morning unless we get into the Columbia. You have to take us in." Shaking his head, Captain Flavel replied, "It is absolutely out of the question. You will have to ride out the storm. Possibly by morning I can take you in. The bar is breaking clear across. You haven't enough steam to cross the bar before dark."

But Captain Thompson insisted, saying that the *General Warren* could "fire up with a lot of bacon fat and dry stuff to make steam. You must take us in." As their exchange became more heated, Captain Flavel argued: "I know the bar. You can never make it. It is suicide to make the attempt." Overhearing his refusal, some passengers joined the argument, "If we could just get hold of a pilot who was not a coward, we could cross the bar." Only 29 years of age, Captain Flavel flushed and said, "A heavy ebb tide is running. It is unsafe, but if you insist, I will take you in, but I will not be responsible for the consequences."

Captain Flavel sent his boat crew back to the *California* with instructions for the pilot boat to follow them in. The *General Warren* crossed the bar just as darkness was falling and it was beginning to snow. The *California* had been unable to follow because the wind had died. With so much water in the hold, the *General Warren* had trouble making headway against the ebb tide and her rudder responded poorly. Not only were the pumps failing to keep up with the leak in the hull, but heavy seas encountered in crossing the bar had shipped so much water over deck and into the hold that it was almost up to the fires in the boilers. The ship had also lost all but one of her small boats lashed on deck. Sensing the inevitable, Captain Thompson said to Captain Flavel, "If we are going to save our passengers, we will have to beach her."

In the darkness and blinding snow, Captain Flavel could not locate Sand Island, so he steered the *General Warren* instead for Clatsop Spit. The wooden bow struck hard on the sands of Clatsop Spit and began to break up almost immediately. Captain Thompson rolled out a barrel of whiskey and told his crew and passengers to help themselves to what might be a last drink. But Captain Flavel intervened, saying, "We need clear minds if we are going to save the men aboard. Help me roll that barrel of whiskey overboard." As the whiskey went over the side, some of the passengers were lighting their pipes for a last smoke while others were praying.

Pointing out the one remaining small boat on deck, Captain Thompson asked Captain Flavel, "Pilot, do you think you can make it? Can you get to Astoria and summon help for us?" In reply, Captain Flavel shook his head and said, "No, we can never live through the breakers, but I am willing to make the attempt." Captain Thompson selected four of his crew and four passengers to man the small boat. As they were readying the craft, Captain Flavel saw a muscular young man who was one of the passengers, J. G. Wall, and asked, "Can you handle an oar?" Wall responded, "I'll go if you need me, though I believe it is sure death."

Surprisingly, when the lashings were cut, the boat launched into 20-foot breakers without capsizing. As they departed, Captain Thompson called out, "Pilot, you will come back?" Captain Flavel shouted back, "If I live, I will return."

Rowing all night through the snow, heavy seas continually filled the boat. She was kept afloat only by constant bailing. Throughout the night, Captain Flavel kept his ears attuned to the roar of the breakers, ordering the oarsmen to row in a direction that kept them away from certain death in the huge waves. When daylight came, they were off Scarborough Head. Sighting a light on Tansy Point, the crew landed where pioneer settler Bartholomew Kindred had built his log cabin five years earlier. Kindred stoked up a roaring fire and brewed hot tea for the crew while his wife Rachael tended the blisters on the hands of some of the oarsmen.

After a quick breakfast, Captain Flavel and his crew rowed another three hours to Astoria. There, they organized a rescue party and secured the assistance of the whaleboat *George and Martha* to go to the aid of the *General Warren*. But when the rescue party reached the spot where the *General Warren* had struck Clatsop Spit, there was almost no trace, only a few scattered pieces of wreckage. Forty-five bodies eventually drifted to shore near the mouth of the Necanicum River, in what is now Seaside, Oregon.

Less than two years later, Captain Flavel was the pilot aboard another vessel lost on the bar. The bark *Oriole,* loaded with a cargo of building materials for a proposed lighthouse at Cape Disappointment, spent 22 days on her voyage from San Francisco before arriving outside the Columbia River on September 19, 1853. Captain Flavel was the pilot and began the crossing with the flood tide and adequate wind. Midway through the crossing, when the wind was lost and the incoming tide continued to push the *Oriole* toward Clatsop Spit, Captain Flavel ordered the anchors dropped, but neither held, and the *Oriole* dragged anchor until she hit Clatsop Spit close to the spot where the *General Warren* met her demise. Captain Flavel and the 32 passengers debarked the *Oriole* into lifeboats and stayed offshore until sunrise when they were rescued by the *California.*

George Flavel was both an outstanding navigator and extremely capable businessman. Under his command, the barkentine *Jane A. Falkenberg* set the speed record for a voyage from Astoria to San Francisco at three and one-half days. This vessel was never placed into pilotage service, but her profits in shipping helped Flavel build a fortune based upon pilotage and multiple business interests in Astoria.

On Christmas day in 1854, the Clatsop County commissioners met in a special session and appointed Captain Flavel harbor master at the Port of Astoria. This position was unsalaried, but gave Flavel the authority to regulate when ships sailed across the bar. Given the relatively shallow depth of the bar at that time, a ship with any significant draft needed to schedule her departure on the flood tide. Captain Flavel's knowledge of the bar, sailing conditions, and his practice of sending experienced pilots to San Francisco to contract pilotage work for voyages to Astoria enabled him to establish a monopoly on pilotage across the Columbia River Bar that prevailed for over 30 years until 1887.

The most serious challenge to Captain Flavel's pilotage monopoly occurred in 1865 when Captain Paul Corno arrived on the bar with his 101-foot steam tug *Rabboni*. "We could put Flavel out of business in no time at all and take over the *California's* trade," bragged Corno to his pilots Andrew Belmont and Moses Rogers. But despite the *Rabboni's* superior speed and power, Corno had underestimated his competition. Captain Flavel maintained his pilot boat *California* outside the river entrance 24 hours a day and continued to send his pilots south to San Francisco to board ships bound for the Columbia River. Despite the *Rabboni's* towing advantage, Captain Corno was unable to overcome the loyalty of West Coast ship captains to Flavel's experienced bar pilots.

Losing money since his first day on the bar on August 3, 1865, a scheme was hatched to drive Flavel out of business and hand control over to Captain Corno and the *Rabboni*. Portland and Willamette Valley shippers who had long complained about Flavel's pilotage rates successfully urged the pilot commissioners to revoke the licenses of the Flavel pilots with the objective of bestowing a pilotage monopoly on Corno's steam tugboat.

The new monopoly lasted just three days as the well-connected Flavel was able to secure licenses for three of his pilots from the county commissioners in Pacific County, just across the river from Astoria, issued under the laws of the State of Washington. Pilots Alfred Crosby, Will Metzger, and Asa Farnsworth were back in business within a matter of hours.

Pacific County's granting of Washington pilot licenses to three Oregon pilots caused a major ruckus in Washington that ultimately led to their revocation, but it was too late for the *Rabboni*. Captain Corno surrendered to Flavel in March, 1886, after just over seven months of a losing operation.

The *Rabboni's* short experience on the bar did have one significant impact: shippers and legislators became convinced that a steam-powered tug would improve pilotage service and reduce rates. In 1868, the Oregon legislature authorized the Pilot Commission to let a contract for a "powerful steam tugboat to be subsidized for a five-year period" for service at the mouth of the Columbia. That legislation required the tugboat contactor to "keep a seaworthy boat of not less than 50 tons burthen, fully decked, to cruise outside the bar of the Columbia River." The subsidy offered by the Pilot Commission was $12,000 for the first year of operations, $6,000 for the second, and $5,000 for the third.

Seeing this offer as an opportunity to diversify his pilotage business and add a powerful tugboat, George Flavel acted quickly to accept the Pilot Commission proposal and within a matter of weeks had contracted for the construction of a new tug in San Francisco at an estimated cost of $40,000. The 101-foot tug *Astoria* was launched in December 1869, and immediately sailed north to Astoria and was put into service on the bar. The first tow was the lumber schooner *Humboldt* on her way to load lumber at Asa Simpson's Knappton sawmill.

Once again, Flavel's shrewd business dealing and timing proved highly profitable. Ship traffic at the mouth of the river increased rapidly, providing plenty of business for both the pilot boat *California* and the tug *Astoria*. Business was so good that Flavel hired three additional pilots. Captains Marshal Staples, Alexander Malcolm, and Erick Johnson brought his pilot orga-

THE CZAR OF THE BAR
Captain George Flavel, circa 1880.

nization to a total of six, and for the first time, he arranged schedules so that no more than two pilots were at sea at any one time.

The colorful "Czar of the Bar" retired from active piloting in 1858, but continued to manage his pilotage business for another 29 years before selling his interest to his longtime partner Asa Simpson in 1887. In more than three decades of managing his pilotage and other business interests, Flavel became the richest man in Astoria, with considerable land holdings, a large dock, warehouses, and his own bank. He was active in politics, regularly entertaining legislators in Salem with free whiskey. He was also elected treasurer of Clatsop County by a large majority and to several terms on the Astoria City Council. Flavel died on July 3, 1893, in his family's mansion, which was the finest home in Astoria when it was built 10 years earlier. The Flavel mansion is preserved as a museum today. Thousands attended his funeral including General J. G. Wall, who had been saved from certain death aboard the *General Warren* over 40 years earlier when Captain Flavel asked the strapping 25-year-old if he could handle an oar in the small boat Flavel launched to seek help for the stranded ship. Wall went on to significant success in the military and in business in northern California, but made his way to Astoria to attend the funeral of his old friend and the man who had saved his life.

The greatest legacy of Captain Flavel's long maritime career was the establishment of a professional pilotage service that was safe, efficient, and founded on the skills of experienced ship captains. The Flavel pilot organization's record of safety was a major contributor to the dynamic growth of both Astoria and Portland during this period. The following excerpt from an editorial in *The Oregonian* in 1875, which ran alongside competing letters to the editor from Portland railroad financier and lawyer William Reid and Captain Flavel, makes exactly this point:

> *Captain Flavel may point with pride to the service which he has rendered this State on the Columbia bar. The record which has been made since the law of 1868 took effect is most emphatically a record of which he and all those connected with him may feel a just pride. Look at the immense shipping that has crossed and recrossed this bar in the care and custody of Capt. Flavel and those connected with the pilotage system of which he is the head; they have successfully piloted hundreds of vessels, from the maximum depth of 23 feet draft down to the little schooner of a hundred tons, and the losses put together has been less than one-sixteenth part of one per cent. Not an accident has occurred to a single vessel having one of these pilots, or one of these tugs, in crossing or recrossing.*

CHAPTER 11

TOLL OF THE BAR

OVER THE LAST 150 YEARS, the Columbia River Bar has claimed the lives of 17 mariners engaged in pilotage service at the mouth of the American West's greatest river. Those who have lost their lives in the line of duty include 12 bar pilots and five pilot boat operators or deckhands who also served as oarsmen in the wooden pulling boats transferring pilots to and from ships in the Pacific Ocean. The circumstances of these passings at sea are briefly described in chronological order:

1855: CAPTAIN JAMES JOHNSON Captain James Johnson was one of three shipmasters sailing under the flag of the Hudson's Bay Company who took over bar pilotage service following the death of Chief Concomly in 1830. In October 1855, Captain Johnson was sailing his homemade sloop across the bar with a neighbor serving as crew when a sudden squall came out of the southwest and capsized the small craft. Neither man's body was ever recovered.

1888: CAPTAIN CHARLES F. JOHNSON On January 10, 1888, heavy fog and high seas made for a difficult boarding by Captain Charles F. Johnson from the pulling boat launched from the pilot boat *Governor Moody* to the British bark *Abercorn*. As the force of the storm increased, the vessel turned seaward to ride it out and avoid the growing breakers extending across the bar. The weather remained thick and foggy for several days. Nothing was heard of the *Abercorn* until three survivors staggered ashore to report the loss of their ship. Captain Johnson was reputed to be a skillful bar pilot, but it was speculated that he lost his bearings in the many days of fog and that the current set the vessel north, sending her ashore 10 miles north of Grays Harbor. When the vessel struck, her mast went overboard almost immediately and heavy surf prevented the launching of the ship's lifeboats. Deeply laden with a cargo of railroad iron, the *Abercorn* went aground so far from the beach that only three survived. Eighteen lost their lives including Captain Johnson.

1890: DECKHAND OLAF HANSEN Olaf Hansen was at work aboard the pilot boat *Governor Moody* when a large wave swept him overboard to his death.

1896: CAPTAIN JOHAN GRASSMAN Captain Johan Grassman boarded the *Cadzow Forest* in a winter storm in January 1896. She was ultimately lost under unknown circumstances and all hands perished.

1900: CAPTAIN PETER CORDINER In the midst of a strong winter storm in December 1890, Captain Peter Cordiner crossed the bar aboard the pilot boat *San Jose* and made a difficult transfer from the pulling boat to the bark *Andrada*. Hoping to ride out the storm and await calmer conditions, Captain Cordiner ordered the *Andrada* to turn seaward, and she then disappeared without a trace. Ten years later, Captain Kevin Kenneally sailed his ship *Jordan Hill* into Astoria and described the plight of the unlucky *Andrada*.

"We were about 20 miles off the entrance and were headed north when we sighted the *Andrada* lying on her beam ends," said Kenneally. "We had been drifting helplessly for two days ourselves with the distress flag flying and we could not go to her aid. There was no sign of life aboard." Captain Cordiner was only 35 years old when the *Andrada* was lost. He left a widow and a baby son.

1911: CAPTAIN THOMAS LATHAM On February 10, 1911, the gas schooner *Oshkosh* departed Tillamook Bay for the Umpqua River in the midst of a rising storm. As the storm worsened, Captain Thomas Latham decided to turn north in the hopes of entering the Columbia River and avoiding the storm. Near the south jetty, huge seas and gale force winds capsized the *Oshkosh,* rolling her over and throwing six of the seven aboard into the ocean. All six drowned including Captain Latham, who was in his mid-50s and had worked his way up from deckhand on George Flavel's tugboat to his bar pilot position. Amazingly, engineer George May, who had locked himself in the engine room, rode the vessel to the beach where he emerged badly bruised, but alive.

1925: CAPTAIN KENNETH P. T. WOOD On December 4, 1925, Captain Kenneth Wood had just finished piloting the steamer *Knoxville City* on an outbound transit and was disembarking from the pilot ladder to the pilot boat when he lost his grip and fell into the sea. According to the autopsy, the fall was the cause of death. Captain Wood was a native of Scotland, 55 years of age and had joined the bar pilots five years earlier in 1920.

1925: DECKHAND PETER ANDERSON Before the year was out, another pulling boat tragedy claimed the life of deckhand Peter Anderson. While rowing the transfer boat between the pilot boat *Columbia* and the steamer *Overjsel,* a rogue breaker capsized the transfer boat and threw Anderson, deckhand William Gunnari, and bar pilot Bob Marquart into the sea. Marquart and Gunnari were able to grab the keel line rigged on the transfer boat precisely for such incidents and were picked up by the steamer, but deckhand Anderson disappeared in the rough seas.

1957: CAPTAIN ELTON GILLETTE Dubbed one of the "Shanghai pilots" because of his time as a pilot in Shanghai, which followed many years as a master aboard States Line vessels engaged in the China trade, Captain Elton Gillette suffered a fatal heart attack on the deck of the pilot boat *Peacock* on February 27, 1957. He was 60 and had served as a bar pilot for over 15 years, joining the organization just before the onset of World War II.

1960: DECKHAND JOHN JENSEN In another pulling boat accident on February 3, 1960, the ship rolled during a pilot transfer and the surge of sea water overturned the boat and oarsman John Jensen was lost. The pilot boat *Peacock* spent two hours searching for Jensen without success, retrieved the pulling boat, and returned to Astoria.

1962: DECKHAND WILLIAM WELLS Deckhand Billy Wells was just 35 when a freak wave caught the pulling boat shortly after it had disembarked bar pilot Edgar Quinn from the freighter *Olympia Maru* on September 28, 1962. Captain Quinn, Wells, and deckhand Don Nelson were able to right the small craft by hauling on her keel line, but then had to bail throughout the night as a huge winter storm drove the boat north. Wells grew weaker as the night progressed and died of hypothermia the next morning. Both Quinn and Nelson survived. Wells' body washed ashore several days later.

1963: CAPTAIN RICHARD J. WILLIAMSON Another heart attack felled a bar pilot with his boots on in 1963. Captain Richard Williamson had just piloted the Texaco tanker *Minnesota* outbound across the bar on March 17, 1963, when he suffered a heart attack on the bridge and died. Just 52 years of age, Captain Williamson served as a master with Dollar Line and American Mail Line before becoming a bar pilot in 1955.

1964: CAPTAIN CARL P. PARKER In his twenty-first year as a bar pilot, Captain Carl Parker was debarking to the pilot boat *Peacock* after having completed an outbound pilotage assignment aboard the *Rachel V* on December 11, 1964, when he became the third bar pilot in seven years to perish from heart failure while on duty. Captain Parker was 59 and had started his long maritime career as a teenager out of Castine, Maine.

1973: CAPTAIN EDGAR QUINN A survivor of the harrowing day and night in a pulling boat in the midst of a massive winter storm in 1962 chronicled in Chapter 8, it was a pilot ladder incident 11 years later that claimed the life of Captain Edgar Quinn. After piloting the freighter *Maritime Queen* outbound on a relatively routine pilotage assignment in moderate seas on March 18, 1973, Captain Quinn was climbing down the pilot ladder to the *Peacock's* waiting daughter boat when the daughter boat caught the ladder and snapped Quinn against the ship's side. He lost his footing, slid down the manropes, and plunged into the ocean. The daughter boat reached him quickly, but neither Captain Quinn nor the deckhand running the daughter boat was able to pull him into the small craft. The Coast Guard helicopter arrived too late to pull him from the sea before he was overtaken by hypothermia. In his distinguished maritime career, Captain Quinn obtained an unlimited master's license at age 27, one of the youngest ever to do so. A veteran of more than two decades with the bar pilots, Captain Quinn lived up to his characterization of the Columbia River Bar as "a place of treachery."

1979: BOAT OPERATOR DAVE ENLUND Boat operator Dave Enlund had worked for the bar pilots for over a decade before suffering a fatal heart attack while onboard the *Peacock* in 1979. Enlund was 52 years old.

2003: CAPTAIN PAUL JACKSON Captain Paul Jackson had been a bar pilot for 17 years when he died of a heart attack on the bridge of the lumber ship *Hoegh Miranda* while he was completing an outbound transit on October 19, 2003. Captain Jackson was typical of the modern bar pilot, graduating from the California Maritime Academy in 1967 and then going to sea for nearly 20 years before joining the bar pilots in 1986. Jackson was civically active in Astoria and only 56 at the time of his passing.

2006: CAPTAIN KEVIN MURRAY A graduate of the Maine Maritime Academy, Captain Kevin Murray had been a bar pilot for just 14 months when he attempted to disembark the log carrier *Dry Beam* to the pilot boat *Chinook* after an outbound transit on January 9, 2006. As he started his leap from the ladder to the deck of the *Chinook*, the pilot boat veered slightly away from the ship's side. Captain Murray attempted to step back to the ladder, but lost his hold on the manropes and slid into the Pacific. The *Chinook* immediately implemented its man overboard procedures, but the search was hampered by the heavy swell and the unfortunate fact that Captain Murray's strobe light did not deploy because the switch had not been turned to the "on" position. The crew of the *Chinook* initially sighted Captain Murray on the starboard bow, but then lost him and did not sight him again for five to 10 minutes. On the second sighting, Captain Murray appeared to be unconscious and an attempt to retrieve him with the rescue basket failed and he was lost again. An intensive search by multiple vessels failed. Captain Murray's body came ashore on Roosevelt Beach, Washington, two days later.

PART THREE

TWO CENTURIES OF BAR PILOT TRANSFER TECHNOLOGY

CHAPTER 12

THE PULLING BOAT ERA

THROUGHOUT THE 200-YEAR HISTORY of commercial pilotage on the Columbia River Bar, it is remarkable that the humble pulling boat or rowboat transferred bar pilots to and from vessels in the Pacific Ocean for nearly 150 of these years. And if one counts Chief Concomly's long dugout canoe, that operated between 1813 and 1830 just inside the bar, small human-powered craft accomplished the pilot transfers for 154 years from 1813 until 1967, when the *Peacock* arrived with her motorized aluminum daughter boat.

During its nearly century-and-a-half of service, the pulling boat was deployed by larger craft first powered by sail, then steam, and finally diesel. But the pulling boat design remained fairly timeless and its use by pilots was consistent up and down the West Coast for over a century. As displayed in the photographs in this chapter, the typical pulling boat was 15 feet, 10 inches in length, and had a beam of 4 feet, 8 inches. Heavily built with oak frames and rot-resistant cedar planking, she had a plumb stem and transom with rising floors amidships. The pulling boat's shape and construction resembled the once-common Whitehall-type ship's launch boat.

Throughout this era, the larger pilot boat carrying both pilot and pulling boat was rigged to carry a pair of pulling boats to facilitate a launch from either the port or starboard side of the pilot boat. Heavy metal eyes were built into each end of the pulling boat for lifting by davits.

Once launched, two oarsmen would row, one in the bow and the second midships, while the bar pilot sat in the stern. Each of the two rowing stations was set up with double thole pins, port and starboard. On the Columbia River Bar, the thole pins were made of wood to facilitate quick breakage of the pins rather than the oars in rough conditions. Extra thole pins were always onboard, lashed underneath the seats. The keel was rounded up at both ends and metal flotation tanks were eventually installed under the seats.

Launching the pulling boat from the davits was relatively easy, but retrieval could be hairy in rough conditions. A surging sea made it especially difficult to place the hooks into the metal eyes at each end of the pulling boat at the same time. If one hook was through but not the other, the pulling boat could become a bucking bronco that was difficult to bring under control to match the second hook with the remaining eye.

In the cold waters of the north Pacific where hypothermia could disable even the strongest swimmer, the wooden pulling boat was the lifeline whenever things went awry and a pilot either ended up in the sea boarding or disembarking or the pulling boat was swamped by a large wave or a surge of sea water from a rolling ship. To assist in righting a pulling boat once she turned turtle, each was fitted with a line threaded back and forth through the keel at 18-inch intervals. The righting maneuver was fairly straightforward, but no easy task: two crew on the same side of the pulling boat grabbed the keel line and pulled her upright.

The pilot transfer process was always the riskiest. Once alongside a ship, the bow oarsman or bowman boated his oars and grabbed the ship's boat rope while the sternman boated his inboard oar and used his outboard oar to maneuver the boat adjacent to the Jacob's ladder. In boarding situations, it was then the bar pilot's task to time his leap to the ladder in boarding an inbound ship. When disembarking a pilot from an outbound assignment, the bar pilot had to time his jump from the Jacob's ladder to the small platform at the stern of the pulling boat with one or two manropes in hand. Whether to use one or two manropes was a matter of personal preference based upon the particular pilot's rappelling technique.

The debarking procedure was usually the most dangerous. Faced with the prospect of a spill into the sea while descending a ladder from a heaving ship next to a bobbing pulling boat, the bar pilot was usually willing to risk injury in order to make certain that he landed in the boat. Twisted or broken ankles were a common occurrence, as were pilots in the water. Captain William Dougan, who served as a bar pilot for 17 years beginning in 1959, earned the nickname "Dunkin Dougan" for his record of six meetings with the sea during pilot transfers.

The pilot boats from the 1850s through the end of the nineteenth century were sailing schooners with two pulling boats each. The pulling boats of that era were built in San Francisco, but Astoria's growth from fishing village to commercial center included a boat-building industry that eventually copied the traditional pulling boat design and construction for the bar pilots. Local builders tried different methods to dampen the heavy pounding that pilot boats encountered in pilot transfers with thick rope bumpers and eventually air-filled rubber buoys, but damage to the relatively soft cedar planking was inevitable. The useful life of a pulling boat was typically no more than eight to ten years. Each summer, the pulling boats were repaired; damaged planking was replaced and the hull refreshed with several coats of white paint. One of the last pulling boats in service on the Columbia River Bar is an exhibit in the Columbia River Maritime Museum. It was built by Astoria Marine Construction sometime in the 1930s.

Steam tugboats operated on the Columbia River Bar from the mid-1860s until the 1920s. Towage was the primary purpose of these vessels, but some were operated by bar pilots or carried pilots, for a fee, to incoming ships. A bar pilot was sometimes transferred by pulling boat to his assigned ship, but it was often the case that he stayed aboard the tug and directed the transit from this lead-in vessel after the towline had been passed to the ship. Before completion of the north and south jetties, tugboats improved the efficiency of navigating the Columbia River Bar by increasing the range of conditions when a bar passage was possible and speeding up the transit for sailing vessels in light wind conditions. The bar pilots continued to use sailing pilot schooners to patrol outside the Columbia River Bar seeking incoming vessels. The sailing schooner was faster and more cost effective in this function than the steam tugboat.

The first steam tugboat was Captain Paul Corno's *Rabboni*, whose pilots failed to make a dent in Captain George Flavel's monopoly. The *Rabboni* operated for less than one year in 1865-66. With the benefit of a state subsidy, Flavel launched the tug *Astoria* in 1869. She was followed by the tug *Columbia*, built by Captain Asa Simpson at his Knappton, Washington sawmill and launched in 1876. She was replaced in 1894 by another Simpson tug, the second *Astoria*.

The tugboat and pilotage business was quite competitive after the death of George Flavel. Over more than four decades between 1889 and 1923, the Union Pacific Railroad, Port of Portland and Puget Sound Tugboat Co. operated tugboat divisions at the mouth of the Columbia River and employed or carried bar pilots. But the fierce competition resulted in losses. Union Pacific quit the tugboat business in 1910 and the Port of Portland's foray into the towboat/pilotage business ended in 1922.

There are no known photographs of the pilot schooners *Mary Taylor*, *California*, *Governor Moody*, and *San*

THE *JOSEPH PULITZER*

A former racing yacht, the *Joseph Pulitzer* served as the bar pilots' pilot boat from 1899 to 1919.

Jose, which operated between 1849 and 1898. Dissatisfied with the slowness and maneuverability of the *San Jose,* the bar pilots emphasized speed in their next pilot boat acquisition. Senior pilot Captain Peter Cordiner traveled by train in November 1893 to New York to examine a number of pilot vessels available for purchase from the New York Pilot Association. His selection was the racing yacht *Joseph Pulitzer.* The vessel had been built four years earlier by Joseph Pulitzer, owner of the *New York Times* and for whom the Pulitzer Prize was named. Pulitzer had raced her twice to Bermuda before she entered the pilotage service in New York Harbor.

After completing the deal, the *Joseph Pulitzer* sailed around Cape Horn and then north to the Columbia River Bar where she went into service in the spring of 1899. As can be seen in the photo on this page, she was a striking ship with sleek lines and served on the Columbia River Bar for nearly 20 years until 1919.

With the retirement of the *Joseph Pulitzer,* the bar pilots made do with a combination of tugboats and temporary pilot boats until 1923 when the former rum runner *King and Winge* sailed into Astoria on September 24, 1923. She went immediately into dry dock at the Wilson Shipyard for an overhaul and a new coat of paint on her wooden hull. Once her shipyard stay was over, Captain Charles Gunderson smashed a bottle of whiskey against her bow and rechristened her *Columbia.*

One hundred feet in length, the *Columbia* was the first dedicated pilot boat on the Columbia River Bar with an engine. But the 140-horsepower engine with which she

THE FIRST PILOT BOAT *COLUMBIA*

The first pilot boat *Columbia,* originally christened the *King and Winge,* operated as the pilot boat from 1923 until 1958. One of her pair of pulling boats is being launched.

arrived on the bar proved inadequate to the conditions and was soon replaced with a 170-horse Atlas-Imperial engine at the Wilson Shipyard. She carried the standard pair of pulling boats, which can be seen in the photograph on this page, one being launched and the other resting in her davits.

Captain Frank Craig, one of the Columbia's regular boat operators, improved the pilot boat launch process with what became known as a Craig hook. Shackled to the ends of the two davit lines and hooked into the metal eyes of each end of the pulling boat, the Craig hook was easily tripped with a quick flip of a lever as soon as the

THE PILOT BOAT *PEACOCK*

The first pilot boat *Peacock* was a 136-foot converted Navy minesweeper. One of her two pulling boats is seen amidships.

THE SECOND PILOT BOAT *COLUMBIA*

The second *Columbia,* launched in 1958, served for 19 years until she was replaced by the third *Columbia* in 1977.

PILOT TRANSPORT VIA ROWBOAT FROM THE *PEACOCK*

In the middle ground, two deckhands row a wooden pulling boat from the first *Peacock* to disembark a pilot who can be seen descending the ship's ladder after directing her navigation on an outbound crossing of the Columbia River Bar. The *Peacock's* second pulling boat remains secure in her davits. Thick rope bumpers helped cushion the cedar craft when alongside a ship.

---

pulling boat hit the water. The process was reversed during retrieval. The invention significantly increased both speed and safety in rough sea conditions.

Eight years into her service on the Columbia River Bar in 1931, a huge wave struck her broadside, carrying away her stack and ripping away her starboard davits, resulting in the loss of one of the *Columbia's* two pulling boats.

The *Columbia* was retired to standby or backup-pilot boat status on July 28, 1947, when a 136-foot converted Navy minesweeper sailed across the Columbia River Bar following her service in World War II. She was promptly rechristened the *Peacock,* taking the name of the U.S. expeditionary sloop that was wrecked in 1841 on the spit that now bears her name. Wooden-hulled like the *Columbia,* the *Peacock* (see page 75) was powered by twin 600-horsepower Cooper-Bessemer engines, making her much faster than the pilot boat she replaced. That speed enabled the *Peacock* to cross the bar in rougher conditions, but she still utilized the pulling boat method to transfer pilots at sea. In another improvement over the *Columbia,* her davit system was stronger with steel structural members.

Thirty-five years after the *King and Winge* became the first of four pilot boats called *Columbia,* steel replaced wood when the newest *Columbia* was launched by builder Gunderson Brothers in Portland in 1958 (see page 75). She was designed to act as a back-up boat for the twice as long and more powerful *Peacock* and to spell the *Peacock* during the good weather summer months that included the *Peacock's* annual multi-week period in dry dock for maintenance and repairs. She was 65 feet in length and powered by two General Motors 12005-A diesel engines that produced a cruising speed of 10 knots.

The 1958 launch of the second *Columbia* marked the beginning of the end of the pulling boat era. Destined for summer service and good weather only, she was not equipped with the standard pair of pulling boats that had dominated pilot transfer on the Columbia River Bar for more than a century. With her steel hull and specially designed rubber bumper encircling the boat at the deck line, she was destined for service in calm to moderate conditions where she could come alongside an incoming or outgoing ship in the Pacific and board or disembark a bar pilot without the need of a small transfer boat.

But for most of the 1960s, the former minesweeper *Peacock* continued to launch pulling boats to accomplish the pilot transfers in the ocean. Sometime in the 1950s, the traditional wooden pulling boat gave way to a fiberglass design with the traditional shape and the addition of an 18-horsepower Evinrude motor. But a pair of oars remained aboard to be unlashed for use when the small outboard failed or was put out of service in a capsizing. It was not until 1967 when this first *Peacock,* which had been in service for 20 years, was replaced by her namesake, a pilot boat with the revolutionary self-righting design and, rather than a pulling boat, a motorized daughter boat launched from a cradle in the new *Peacock's* stern.

Any history of the bar pilots' pulling boat era must mention Don Nelson, who crossed the bar more than 50,000 times in a career that spanned 45 years between 1947 and 1992, with an interruption for service during the Korean War. No one has ever approached Nelson's record 50,000 bar crossings and probably never will, but this revered seaman, often called "Admiral Nelson," had a remarkably charmed career on the Columbia River Bar. While wielding what he called "idiot sticks" in the pulling boats over two decades, he broke his back twice, and his stamina and rowing skill were critical contributors to his survival and that of Captain Edgar Quinn during a winter storm calamity recounted in Chapter 8. Don Nelson was only too happy to retire his oars and pursue the last 15 years of his career as an operator of the new *Peacock.*

CHAPTER 13

THE INCOMPARABLE *PEACOCK* ERA

1967 WAS A YEAR OF MOMENTOUS CHANGE for the mouth of the Columbia River. For more than 150 years, the most humble of vessels – canoes and rowboats – carried pilots to and from their pilotage assignments. In the early 1800s, when Chinook Chief Concomly served as the Hudson Bay Company's first chief pilot, he used a long dugout canoe powered by as many as 20 Indian paddlers to meet a ship near the river's mouth and then guide her through the shifting channels around Middle Sands.

After Concomly, whether meeting an incoming ship offshore or returning home after an outbound assignment, the bar pilots covered most of the distance in a pilot boat of sufficient size to carry several pilots and to provision both pilots and crew for several days at sea. Between the 1830s and the 1960s, pilot boats progressed from sailing schooner to steam-powered tugboat and then diesel-powered pilot boats, but these vessels did not actually perform that last critical task of putting the pilot aboard the incoming ship or disembarking the pilot upon completing an assignment. That function was left to a sturdy wooden rowboat, or pulling boat, powered by a pair of oarsmen with the pilot seated at the stern.

Throughout most of this era, the world's fleet was growing in average vessel size, but not dramatically so. The larger cargo vessels of the late nineteenth century approached 300 feet in length, had a beam of 45 feet, and carried a crew of 36 to 40. Nearly 50 years later in the 1930s and 1940s, cargo ships of the size of the *Iowa* were common, with a length of some 400 feet, a beam of 50 feet, and a crew of 36. Throughout the world, the rowboat continued to handle most pilot transfers, whether boarding or disembarking.

But after World War II, as the U.S. and world economies expanded, a new generation of cargo ships exploded in size, dwarfing their seagoing predecessors. In 1956, the first containerized cargo vessel was launched, and this began a rapid transformation of ocean carriage from the traditional breakbulk vessel that required significant time in port to fit a cargo of all manner of shapes and sizes into multiple holds. Container ships with standardized containers could be loaded quickly and facilitated intermodal transportation. The container loaded in the hinterlands of Europe traveled by truck for loading aboard a container ship in Rotterdam, sailed to New York, was discharged to a train bound for Chicago, and was then trucked to its ultimate destination outside that city.

As ships grew in size, and bulkers earned the nickname "Panamax," meaning the vessel was such a behemoth that it maxed out the dimensions of the locks in the Panama Canal, the rowboat era of pilot transfer throughout the world began to wane. It was simply too slow and increasingly difficult for a pair of oarsmen to accomplish the pilot transfer with any speed in moderate conditions. Small boat pilot transfer was also complicated by ladders that now extended several stories from deck to sea and would swing outboard of the rowboat with just a moderate roll of the ship.

As the versatile and relatively small-sized breakbulk fleet gave way to large specialized container ships, oil

Comparison of world's fleet, past and present

***Columbia Rediviva*, 1787** Length: 84 feet Beam: 24 feet Draft: 11 feet

***Peter Iredale*, 1890** Length: 287 feet Beam: 40 feet Draft: 23 feet

Liberty ship, 1943 Length: 441 feet Beam: 56 feet Draft: 28 feet

Panamax, in use today Length: 965 feet Beam: 106 feet Draft: 43 feet

Post-panamax, in use today Length: 1,200 feet Beam: 160 feet Draft: 50 feet

The huge change in the size of the world's shipping fleet during the last 250 years is shown to scale above. With its navigation channel deepened to 43 feet in 2010, Columbia River ports see Panamax and larger vessels up to 1,000 feet in length. The channel will have to be deepened further to accommodate the largest post-Panamax container ships, which reach 1,200 feet in length and draw 50 feet.

tankers, and bulk carriers, the speed and reliability of pilotage service became increasingly important. Shipowners now had such a massive investment in these ships that tight schedules became the norm. The better the management of a ship and her schedule, the more freight-paying voyages and the greater the gross revenue in any year. As a result, pilot organizations throughout the world shifted from human-powered rowboats to a wide variety of powered pilot boats appropriate to the conditions on their pilotage ground.

At the mouth of the Columbia River, the bar pilots knew that their pilot transfer system had to change, but they faced a much greater challenge. Especially during the winter, the bar was closed 20 to 30 times annually when swells reached 15 feet or more and winds exceeded 35 to 40 knots. In these conditions, it was not safe to launch and row a 16-foot rowboat to transfer or disembark a pilot in the open ocean. With the new focus on efficient ship scheduling that was growing each year in the second half of the twentieth century, it was clear that the Columbia River would be at a major disadvantage in attracting the widest range of cargoes if transiting the bar earned a reputation as "highly unpredictable" throughout the world shipping community. As world shipping increased throughout this period, competition among West Coast ports was keen: any disadvantage would be exploited and the capital necessary to fund large and expensive infrastructure at both private and public terminals would be in jeopardy. The Port of Seattle started running ads: "We stay open year round, unlike the Columbia River."

As former shipmasters, some with time ashore involved in fleet management, the bar pilots were keenly aware of the threat that modern shipping trends posed to the competitive health of the Columbia River transportation system. In the early 1960s, there were two pilot boats in service. The 65-foot *Columbia,* which had just been delivered in 1958, served as the bar pilots' summer boat. The 136-foot *Peacock,* a converted Navy mine sweeper, was the heavy weather vessel, typically staying on station from sometime in September to sometime in May. The *Peacock* was capable of staying outside in the Pacific during all but the most extreme winter weather; in fact, it was the transfer pulling boat that was the weak link in the system. The bar pilots simply had to find a way to overcome the inability of the oar-driven or small, outboard-engine-powered rowboat to transfer and disembark pilots in sea conditions of greater than 15 feet.

In the mid-1960s, a number of bar pilots hatched the idea of building a new pilot boat modeled after the German North Sea rescue cruisers. The latest line of these rescue vessels was double-hulled for extra strength, self-righting, and carried a 23-foot "daughter boat" launched from a sloped ramp incorporated into the hull. Powered by a 105-horsepower Mercedes engine, the daughter boat could make 12 knots and, the bar pilots believed, would enable them to transfer pilots to and from ships in seas of 15 to 25 feet. If so, bar closures could be reduced to a minimum and passage to and from Columbia River ports would be as regular and consistent as anywhere in the world.

Over the course of three years, with Captain Kenneth McAlpin as the lead representative, the bar pilots worked with Maierform of Bremen and Geneva, the designer of the latest in the long line of rescue vessels built for the German Lifeboat Institution. The prototype was the rescue cruiser *Arwed Emminghaus,* but the bar pilots insisted on significant modifications to reconfigure the *Emminghaus* class of rescue cruiser for the nasty pounding a pilot boat would frequently encounter on Columbia River Bar. The bar pilots specified that the hull be lengthened by four feet and that the interior watertight bulkhead spacing be doubled from a single compartment to a two-compartment design. The *Emminghaus* class of vessel was double-hulled for extra strength, but a breach of her single compartment could sink her. With a two-compartment design, one compartment could be flooded and the vessel would still retain her buoyancy and stability. The new pilot boat retained the rescue cruiser's curved deck design of extreme camber or "whaleback" design for the most efficient shedding of water over the deck. The vessel was also self-righting, meaning that she could be flipped over by a huge breaker or "turn turtle," pop upright to the surface, and keep on running.

Once the bar pilots and Maierform reached agreement on the new *Peacock's* design, she was built in 1966–67 by Schiffs and Bootswerft at the Schweers Shipyard in Bremen, Germany. As built, the steel

THE DOUBLE-HULLED FOREBODY

To withstand the pounding of winter storms, the *Peacock* was double hulled. This photo shows the longitudinal and web frames in the pilot boat forebody.

SETTING THE WHEELHOUSE

The *Peacock's* wheelhouse is lifted into place during construction in Bremen, Germany.

DROPPING THE ENGINE

Installation of the *Peacock's* 1,350-horsepower center engine.

DELIVERY OF THE *PEACOCK*

This *Oregonian* photograph captures the offloading of the *Peacock* and her daughter boat from the German freighter *Ostfriesland* July 31, 1967.

RETRIEVING THE DAUGHTER BOAT

The daughter boat being retrieved and winched up the rollers in the *Peacock* stern ramp with boat operator Bud Plummer at the controls and bar pilot Captain Paul Jackson aboard.

THE *PEACOCK* AT WORK

The *Peacock* underway in moderate conditions in the Columbia River.

double-hulled vessel displaced 72.5 tons. She had an overall length of 87.9 feet, a beam of 18.4 feet, and a draft of 5.6 feet. She was powered by three Mercedes engines, a 1350-horsepower center engine, and two 525-horsepower side engines. The 23-foot aluminum hull daughter boat displaced 3.15 tons and was fitted with a 105-horsepower Mercedes engine.

During sea trials in Germany, the *Peacock* averaged 22.5 knots in three runs and the daughter boat made 12.4 knots. This was slightly below the contract specifications of 24 and 14 knots respectively, but this performance drew no objections from the three bar pilots in attendance for the sea trials, Captain Kenneth McAlpin, Captain Edgar Quinn, and Captain Bill Dougan. In terms of construction costs, the German shipyard produced final bills very close to their original quotes. The *Peacock* quote was $450,000 and the final bill tallied $459,293. The daughter boat was quoted at $30,000 and came in at $31,793. The *Peacock* and daughter boat were loaded aboard the German freighter *Ostfriesland* in July and made the transit from Germany through the Panama Canal and on to the mouth of the Columbia River. On July 31, 1967, Captain Martin West boarded the *Ostfriesland,* transited the bar, and directed the anchoring of the vessel above Buoy 40, just off the current location of the Columbia River Maritime Museum.

As delivered, the *Peacock* was radically different from any pilot boat in the world. She was the only self-righting pilot boat anywhere and her daughter boat was intended to board and disembark pilots in seas of 15 to 25 feet, conditions that shut down pilot service everywhere else. The *Peacock* even looked radical for a pilot boat: she was soon dubbed "submarine with a bow."

Once christened and launched, it took more than a year for the bar pilots to fully utilize the design capability of the *Peacock* and her daughter boat system. But as planned, she revolutionized the availability of pilotage service during heavy weather conditions on the

NEGOTIATING A TRANSFER

Bar pilot, boat operator and *Peacock* daughter boat approach a ship in moderate seas. At the crest of the swell, the bar pilot jumps to the ship's ladder for a pilotage assignment and the daughter boat turns away to return to the *Peacock* for retrieval.

Columbia River Bar. Bar closures per year plummeted to relative insignificance. The daughter boat system proved itself as each daughter boat operator and bar pilot learned how to exploit the craft's speed and maneuverability to approach a ship in heavy seas and time the pilot's jump to or from the ladder in seas of 20 to 25 feet. Over the course of a remarkable 32 years of service on the Columbia River Bar, many a bar pilot who boarded an incoming ship in such extreme conditions was greeted by the ship's captain on the bridge with the exclamation, "There's no place else in the world where I could expect to get a pilot in these conditions."

The courage and skill of each bar pilot in pushing the daughter boat to her design limits in rough conditions was extraordinary. In heavy seas, while holding on to a line tethered to the side of the daughter boat closest to the ship to be boarded, the bar pilot had to time that leap to the ladder at the crest of the swell. Leaping too early could expose the pilot to being crushed by the daughter boat as she continued to ride up the swell and potentially collided with the side of the ship. Disembarking was even more dangerous. Depending on their personal preference, a bar pilot used one or two manropes to rappel the last 10 to 15 feet from the ladder to the daughter boat in a surging sea.

One might ask how is it that the *Peacock* was able to make such an advance over the system she replaced, adding 10 feet of nasty sea state to the capability of the pilotage system on the Columbia River Bar? The explanation can be found in her unique but sturdy design and the commitment of the bar pilots to embracing that design and pushing it to its physical limits.

Sporting an orange, white, and green paint job, the *Peacock* had three staterooms below with sleeping accommodations for 11 including three crew and up to eight pilots. In times of heavy ship traffic, the *Peacock* was filled with as many as eight pilots coming, going, or waiting for their assignments. The *Peacock* had a fully equipped galley and fuel tank capacity of more than 4,000 gallons. At full cruising speed of 14 to 16 knots, she burned approximately 1,000 gallons every four days.

Given the dangerous and tricky character of launching a small boat from a mother craft in rough sea conditions, the *Peacock* was designed to reduce the hazards. A deep, V-shaped launching well or cradle, slanting down to the sea, made up most of the aft third of the *Peacock's* steel hull. At the *Peacock's* stern, a watertight, hydraulically operated ramp was lowered whenever the crew prepared to launch or recover the daughter boat.

When actually launching, a crewman and a bar pilot climbed aboard the daughter boat while it remained in the cradle and lowered the stern ramp. The boat operator running the *Peacock* maneuvered her into a lee created by the incoming or outgoing vessel. When the *Peacock* was in position, the bar pilot climbed aboard the daughter boat, took a place in the bow and pulled the line releasing the pelican hook holding the daughter boat in her cradle. She then slid down a set of rollers and launched into the sea astern of the *Peacock*.

During recovery, this process was reversed, but the approach from the sea into the cradle required a fair bit of skill. The daughter boat operator had to hit the center of the cradle at a full speed in order to carry enough momentum to slide over the rollers up into the well where the boat operator quickly snapped the towline into place and then activated a winch to pull the daughter boat the remaining distance into her nesting position. The watertight stern ramp was then hydraulically lifted back into place. Regardless of the weather, the recovery operation rarely took more than three minutes, but required carefully choreographed maneuvers by the daughter boat operator and the boat operator running the *Peacock*.

Despite the pains taken by the bar pilots during the design process to overbuild the *Peacock* for the conditions at the mouth of the Columbia River, modifications were nonetheless necessary to keep her fit for her intended heavy weather service. The German Rescue Service practice of not attempting to recover the daughter boat until both vessels were in calm or protected waters was not practical for an efficient pilotage service where ships often had to be worked in succession and daughter boat retrievals had to occur quickly. It soon became apparent that the *Peacock's* ramp and stern design was inadequate because the ramp hinges and hydraulic ramps on the articulating stern were too weak to withstand the pounding when the daughter boat was driven up that ramp and into the cradle.

During her first winter of service, the stern ramp was braced and welded open as a temporary fix. The following spring, when the summer vessel *Columbia* moved into service, the *Peacock* was hauled out at Astoria Marine Construction Company. Under the watchful and creative eye of yard supervisor Don Fastabend, the stern ramp and its hydraulics were beefed up substantially, a fix that was good enough to last for another three decades.

The *Peacock's* original Mercedes engines were difficult to start, requiring 15 minutes of careful attention to a detailed checklist. The complexity of the boat's three controllable pitch propellers required considerable experience to master. Both the Mercedes engines and controllable pitch propellers were eventually replaced with American engines and fixed pitch propellers.

The other half of the *Peacock* story was how the bar pilots pushed both mother and daughter boat to their design limits and accomplished the extraordinary feat of providing pilotage service to ships in seas of up to 25 feet. Over more than 30 years, the *Peacock* attained legendary status on the West Coast. Two incidents exemplify her unsinkable reputation.

Just over three years following her delivery, on December 7, 1970, a huge breaker completely engulfed the *Peacock*, turning her sideways with such force that the daughter boat was ejected out of her cradle. The *Peacock* was fully broached and laying her port side under water, but then popped back to the surface. Cap-

NOT SOMETHING YOU SEE EVERYDAY

The *Peacock* being lifted ashore in 2010 in route to her exhibit at the Columbia River Maritime Museum in Astoria, Oregon. *Photo by Robert Johnson.*

tain Edgar Quinn, the number one pilot aboard at the time, asked boat operator Ted Mather, "Mr. Mather, are the engines still running?" "Yes, sir," Mather replied. Captain Quinn then ordered, "Would you please get us the hell out of here," and the boat operator promptly complied. The Coast Guardsman on duty at Cape Disappointment saw the huge wave completely envelop the *Peacock* and assumed he was about to make a radio call to send both rescue helicopter and vessel to look for the pilots and crew who might have survived the *Peacock* sinking. Instead, it was only the daughter boat that sank and she was eventually replaced by a pair of daughter boats designed by Portland, Oregon marine architect Don Hudson. Having a spare proved important as a second daughter boat was lost in 1971.

The second legend-making incident proving the mettle of the *Peacock's* unique design occurred in 1980s when boat operator Don Nelson was attempting to bring the *Peacock* back into Astoria. In the vicinity of Buoy 2, a huge wave picked her up and laid her on her port side. But rather than turning turtle, the *Peacock* surfed down the face of the breaking wave on her beam ends for what

THE 1977 PILOT BOAT *COLUMBIA*

The third pilot boat *Columbia* was launched in 1977 to serve as the fair weather partner of the *Peacock*. While not self-righting like the *Peacock*, she was equipped with a daughter boat launched and retrieved via a hinged stern.

Nelson described as an eternity. When the wave finally passed underneath them, the *Peacock* had been carried a distance of two nautical miles and Captain Nelson, his crew, and several bar pilots found themselves at Buoy 6. It is precisely these conditions – where a pilot boat or ship can be lost – that make the Columbia River Bar so treacherous to transit.

In 1977, the bar pilots commissioned a third *Columbia* to replace her 1958 namesake and to serve as the fair weather partner of the *Peacock*. Built by Nichols Brothers Boat Builders in Freeland, Washington, she was steel-hulled, 82 feet in length and had a beam of 23.5 feet. Like the *Peacock*, she was designed with a hinged stern to accommodate a welded aluminum 25-foot daughter boat. Retired in 2007, her service on the Columbia River Bar spanned three decades.

Launched in 1967, the *Peacock* was finally retired from active service 32 years later in late 1999. Her revolutionary design and the courage and skill of the bar pilots in using the unique capability of her daughter boat system to provide pilotage in that unheard of zone of 15- to 25-foot seas played a critical role in the development of the Columbia River into one of the largest export gateways in the United States, ranking second behind New Orleans in total bulk cargo exports.

If the *Peacock* had to be replaced, it was fitting indeed that her substitute would not be just another vessel, but the combination of the first-ever helicopter engaged in U.S. pilotage and a fast aluminum jet-powered pilot boat that would set the standard for a new generation of fast-run pilot boats throughout the world.

It is also fitting that the *Peacock* now rests ashore in an exhibit at the Columbia River Maritime Museum. Upon her dedication, the *Peacock* became the second of two historic monuments to an extraordinary class of vessel – the famous Theodor Heuss class of German rescue cruiser. With the *Peacock* on display in Astoria and one of her sister ships at the Deutsches Museum in Munich, the pair are bookend reminders of a European ship design legendary for its service in the worst of conditions.

OT BOAT *PEACOCK*

; a one-of-a-kind pilot boat when she began her service on the Columbia River Bar in 1967. She was the only
ot boat in the world and had a unique hinged stern design that allowed the launching and recovery of a smaller
to transfer bar pilots to and from ships in the Pacific Ocean. Based on a North Sea rescue boat design and built in
acock remained in service for 32 years. She was decommissioned in 1999 after having crossed the Columbia River Bar
)0 times. In 2010, the *Peacock* was placed on permanent display outside the Columbia River Maritime Museum in
.

ed to stay at sea for
ned a full galley and
v deck. There was
l up to eight bar pilots.

The *Peacock* was designed to be self-righting. If she was flipped over by a wave while working the dangerous water of the Columbia River Bar, she would continue to roll to an upright position.

The steel, double-hulled *Peacock* was 88-feet long, weighed 100 tons and had a maximum speed of 16 knots. The aluminum-hulled daughter boat was 23-feet long, and powered by a single 105-horsepower engine.

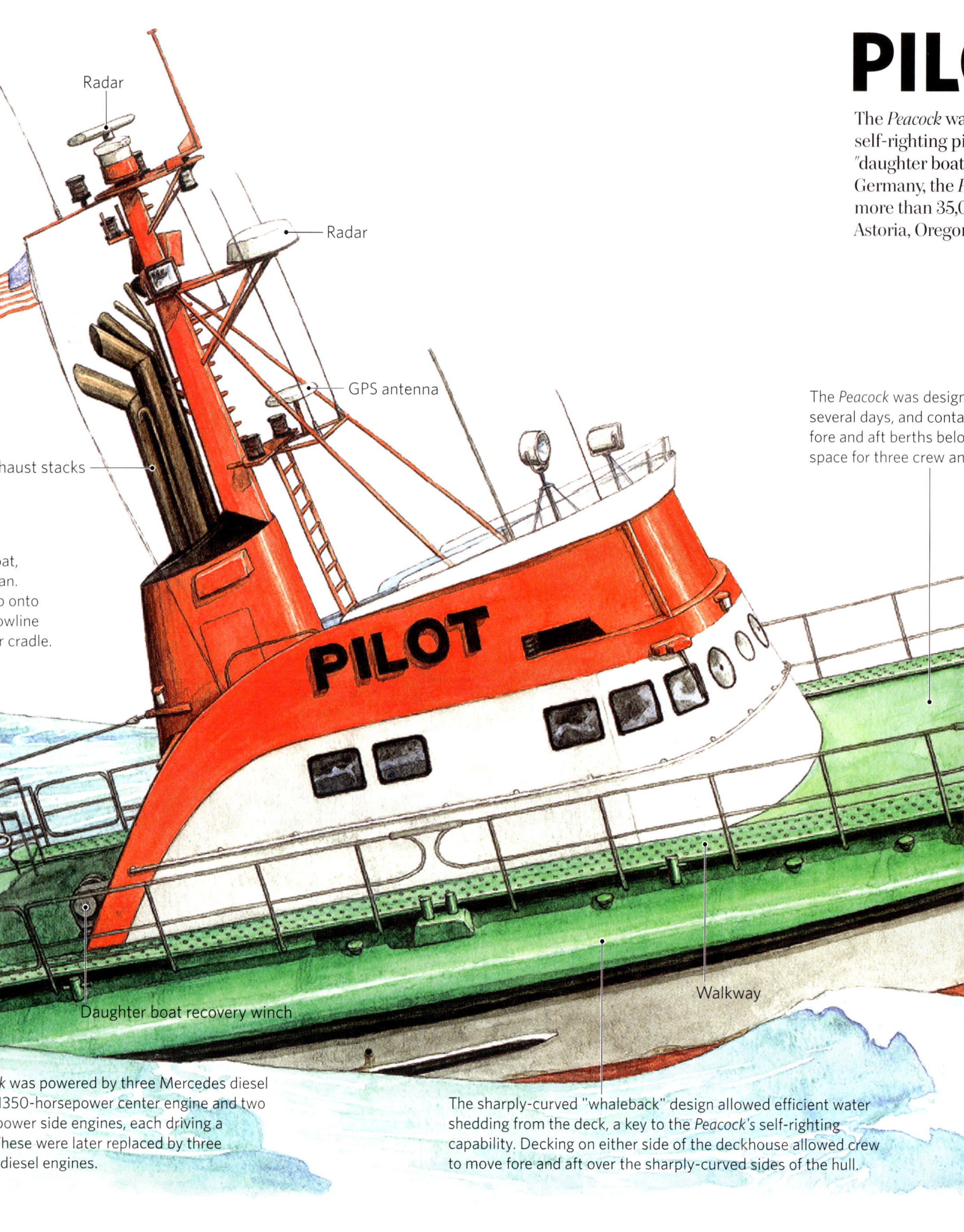

PIL

The *Peacock* wa
self-righting pi
"daughter boat"
Germany, the *F*
more than 35,0
Astoria, Oregon

The *Peacock* was design
several days, and conta
fore and aft berths belo
space for three crew an

oat,
an.
o onto
owline
r cradle.

k was powered by three Mercedes diesel
1350-horsepower center engine and two
oower side engines, each driving a
These were later replaced by three
diesel engines.

The sharply-curved "whaleback" design allowed efficient water shedding from the deck, a key to the *Peacock's* self-righting capability. Decking on either side of the deckhouse allowed crew to move fore and aft over the sharply-curved sides of the hull.

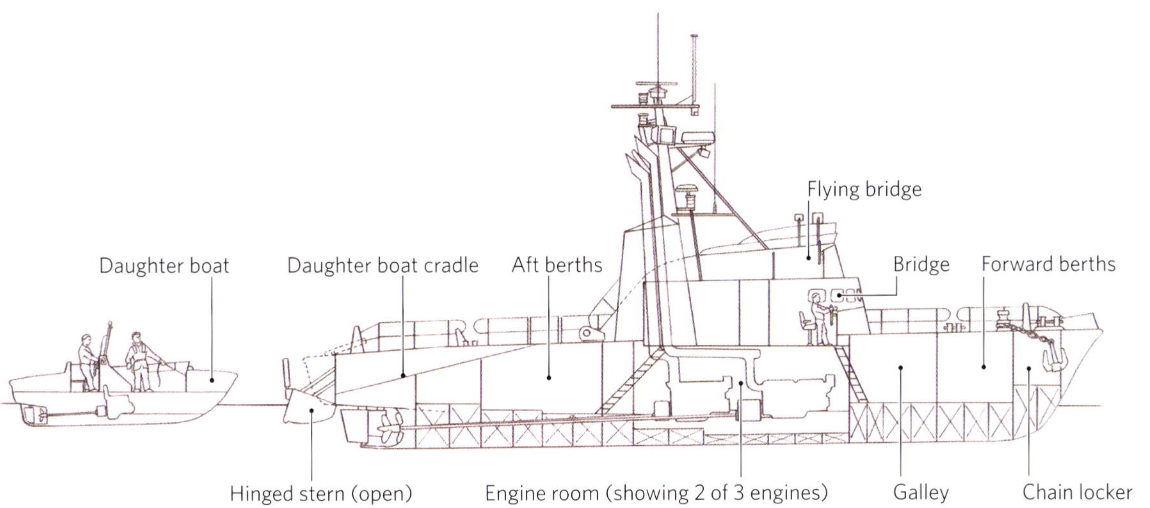

A hinged stern on the *Peacock* would open to allow the daughter boat to slide on and off. Partly submerged when open, the hinged stern had a series of rollers to guide the daughter boat during launch and recovery.

A pull on a cable tether would release the daughter b
allowing her to slide out of the cradle and into the oc
Upon her return, the daughter boat would be driven u
the hinged stern. A crewman would quickly attach a
and the daughter boat would be winched back into h

The *Peaco*
engines; a
525-horse
propeller.
Caterpilla

CHAPTER 14

THE MODERN HELO/FAST BOAT COMBINATION

A PILOT BOAT TYPICALLY HAS A USEFUL LIFE of 15 to 20 years. Given the extraordinary success of the *Peacock* in her first 15 years of service that began in 1967, one would have expected the bar pilots to order a replacement *Peacock* that was essentially an upgraded model with some design changes reflecting years of experience with a one-of-a-kind design. And that was the plan for nearly two decades. The bar pilots had every intention of returning to Maierform's German shipyard for a new and improved *Peacock* with her daughter boat system.

By the mid-1980s, however, questions began to arise as the daughter boat boarding and disembarking system encountered difficulty transferring pilots to and from a particular category of vessel, those with tall sides or high windage. Windage is a term describing the sail effect of the side of a ship above the waterline. As modern cargo ships became behemoths, often exceeding three football fields in length, and rising five to seven stories above the sea to the deck and then another several stories in above-deck superstructure, the force of the wind on the windward side of the ship became a major consideration in boarding and debarking pilots. As ships grew, the windage factor became significant not only for container and car ships, but also for large bulk carriers.

In heavy seas outside the bar, the daughter boat could not simply run up alongside a vessel in the hope that the pilot could grab a heavy ladder swinging back and forth in the wind and sea. In these conditions, the pilot must instruct the ship's captain on how to position the vessel with one side into the wind and the other side creating a sheltered lee to facilitate the approach of the daughter boat. With the wind and sea effects dampened somewhat on the leeward side of the ship, the boat operator and pilot could focus on maneuvering among the swells to make the approach to the pilot ladder. But as the windage factor increased with the galloping growth of ship sizes, there were more and more winter situations where the windage effect on the vessel was so significant that the ship could move sideways across the water at speeds that put the daughter boat at risk of being run over as she attempted to maneuver close to the ship's ladder.

As the height of ships grew, the impact of a vessel's roll in heavy seas dramatically increased the potential outward swing of the pilot's ladder, sometimes sending it beyond the eight-foot width of the daughter boat. The illustration on the next page shows that just a 15-degree roll of one of the average bulk carriers of the 1980s and 1990s, a Panamax ship, would put the lower end of the pilot ladder several feet beyond the outer edge of the *Peacock's* daughter boat. The largest vessels calling on the Columbia in the 1990s could send the ladder more than twice the daughter boat's beam, as shown in the chart. It is worth noting that for the smallest vessel displayed, which depicts one of the largest vessels calling on the Columbia River when the *Peacock* was first delivered in 1967, a 15-degree roll still put the ladder over the deck of the daughter boat.

By the mid-1990s, everyone expected that the trend toward increased vessel size would only accelerate into the next century. Indeed, it was this consensus of where oceangoing shipping was headed that caused

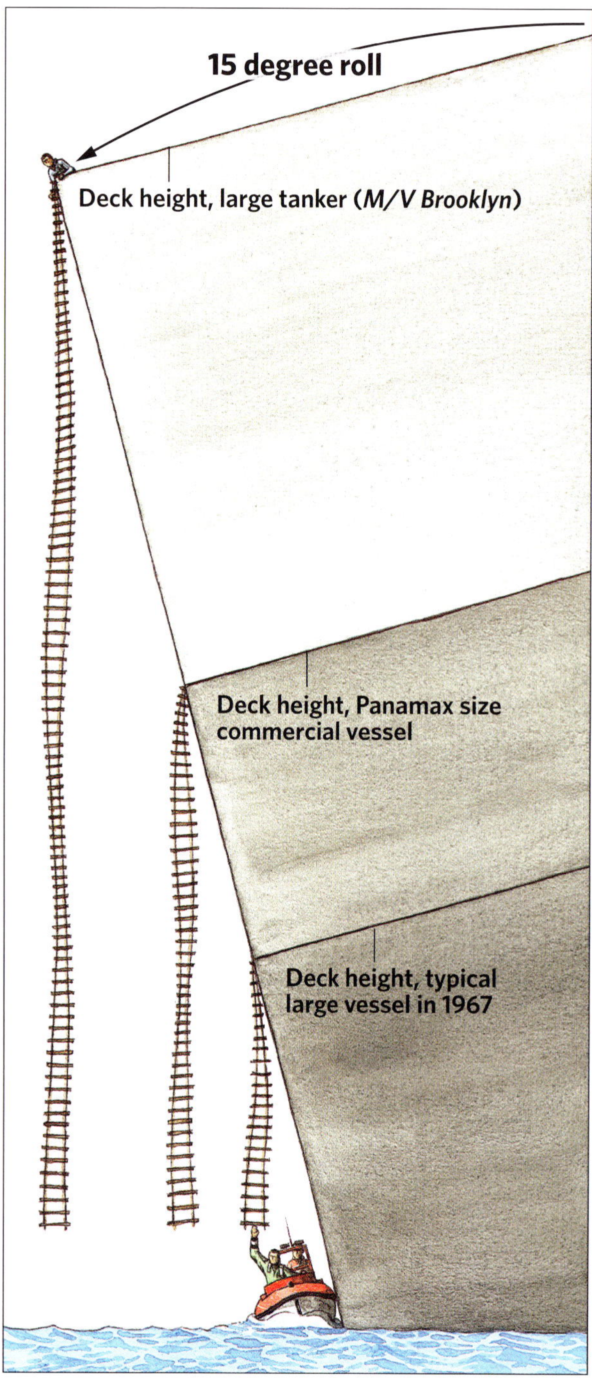

THE DANGER OF SHIP'S ROLL

As ships grew larger, the *Peacock's* daughter boat boarding system became increasingly dangerous. The larger the ship, the greater the swing of the ship's ladder and the greater the risk of injury to the bar pilot.

the U.S. Army Corps of Engineers in 1996 to commence a multi-year study of the feasibility of deepening the Columbia River channel from 40 to 43 feet. The Corps' final 1998 feasibility report confirmed the trend toward larger vessel size and concluded that deepening the river channel from Astoria to Portland was a good taxpayer investment, generating $3 for every $1 of a contemplated $200 million dollar outlay. The deepening project was completed in 2010.

The bar pilots' concerns about the long-term viability of the *Peacock* daughter boat system were heightened by three incidents where a bar pilot was lost overboard while boarding or disembarking in heavy seas. In 1988, Captain Gene Nemecek fell into the sea when a pilot ladder broke. He hit the water tangled up in the ladder and was jerked free, but then had to push off the side of the ship to clear the huge propeller and certain death. This happened so suddenly that the daughter boat lost sight of Captain Nemecek and actually hit him with its propeller in searching for him. The propeller raked down his back but fortunately damaged only his thick float coat. This led to the placement of cages around the daughter boat propellers to prevent any future injury in a similar incident.

In 1994 and again in 1998, Captain Mike Dillon fell into the sea while trying to disembark a high windage vessel in rough conditions. These conditions and the rolling over of the vessel prevented the pilot boat from holding its position as Captain Dillon made his final jump for the deck from an outbound ship at night and fell into the sea. In the 1994 incident, the pilot boat could not find him. Conditions were so treacherous that the Coast Guard would not allow its heaviest weather rescue craft to cross the bar. It was ultimately a Coast Guard rescue helicopter that miraculously spotted the reflective tape on Captain Dillon's life vest and rescued him before exposure caused the first bar pilot fatality since Captain Quinn in 1973. In the 1998 incident, Captain Dillon was rescued by the pilot boat within minutes, but not before encountering the scare of having to swim hard to avoid being sucked into the ship's propeller.

The bar pilots also became aware that the *Peacock's* construction in Germany utilized a type of insulation that was no longer in use and produced toxic fumes in a fire. Any fire aboard the *Peacock* would quickly kill anyone who was not able to vacate the cabin or the bridge very quickly. The three pilot overboard incidents, the flammable insulation, and the ever-growing size of the average ship led the bar pilots to reexamine every assumption about their system and to scour the world for advances in pilot transfer system technology to maintain and hopefully build upon the success of the *Peacock's* now outmoded daughter boat system.

Beginning in 1986, the bar pilots began to assess the potential for a helicopter to become the primary means of moving pilots to and from ships throughout the year and especially in bad weather. The use of helicopters for pilot transport was not new. In the Netherlands' port of Rotterdam, the world's busiest, the use of a helicopter was first introduced in 1965, two years before the delivery of the *Peacock*. Over the next two decades, helicopter service became an integral part of the pilotage system in four more European ports and another four in Australia and New Zealand.

Over the next 10 years, the bar pilots visited pilot groups throughout the world with dedicated helicopter service, observed their operations, and tried out being winched by helicopter cable to and from a ship's deck. By the mid-90s, the helicopter was the primary means of transporting pilots in many European ports. It was safe, minimized pilot traveling time, guaranteed pilot availability in nearly all weather conditions, and promoted fast service for vessels dependent on tight liner scheduling.

But the feature of helicopter service that sparked the greatest interest among the bar pilots was its nearly spotless safety record. In Hamburg, Germany, 25,000 transfers had been accomplished without accident since helicopter service was introduced in 1975. The notion of a virtually injury-free work environment was almost unfathomable to the bar pilots. For over 150 years, the Columbia River Bar Pilots had a saying about the risk of injury in their profession: "Bar pilots don't retire; the Bar retires pilots." In making that jump to or from a moving daughter boat or pilot boat, serious knee and other injuries were common. Indeed, serious injuries were such a significant job risk that their likelihood had to be covered through a combination of a generous sick leave policy and long-term disability insurance.

For an organization steeled to the tradition of boarding or disembarking via pilot ladder in seas of 20 feet or more, the European practice of suspending pilot boat service whenever seas reached nine feet and then switching entirely to helicopter service was a revolutionary concept. But could it work on the Columbia River Bar?

The near fatal loss of Captain Mike Dillon in a pilot ladder accident in 1994 was a wake-up call for the bar pilots' organization and led to negotiations with the Columbia River Steamship Operators Association to test the use of helicopter service. The CRSOA agreed that the bar pilots could borrow the money to fund a four-month test. During the last three months of 1996 and January of 1997, the bar pilots contracted with Petroleum Helicopters to test the use of helicopter transport on the Columbia River Bar. The results were impressive. During this experiment, the helicopter served 344 ships with an average transit time per ship of approximately 20 minutes during daylight hours and 27 minutes during night hours, compared to an average of two hours transferring a pilot by pilot boat.

The ability of the helicopter to transport a pilot so quickly to a vessel offshore enabled that pilot to complete a pilotage assignment much more efficiently than with the use of a pilot boat. With the pilot boat, an inbound pilotage assignment generally took five to eight hours. With the helicopter, the same pilot could handle two assignments in the same time frame.

But despite its remarkable capability in rough weather, there were other conditions – all related to minimum standards of visibility – where the helicopter could not fly. At the mouth of the Columbia River, these included mainly summer days when the fog rolled in and days throughout the year when a low ceiling or other conditions dropped visibility below the FAA-mandated minimum of one mile. In every part of the world where helicopter service had been introduced, the pilotage system still paired a pilot boat with the helicopter. But what type of pilot boat was now optimal for the Columbia River Bar?

Again, after evaluating every option and observing the newest generation of pilot boats in service throughout the world, the bar pilots ultimately opted for boat speed and maneuverability to complement the helicopter's major speed advantage. The field was narrowed to a pair of candidates. One was a 60-foot fiberglass vessel manufactured in the United Kingdom by Halmatic Ltd. Hundreds of Halmatic pilot boats were in service on pilotage grounds throughout the world. The particular design being considered had performed well in the North Sea, where winter conditions approach those commonly found on the Columbia River Bar.

The second design under active consideration was that of a Scottish naval architecture firm, Camarc Ltd. At 75 feet, it was much smaller than the *Peacock,* had an aluminum hull, and was powered by jet engines that could reach speeds of 30 knots.

After the four-month test of the helicopter that ended in early 1997, it was clear to the bar pilots' transportation committee and its top leadership that the combination of helicopter service and a high-speed, maneuverable pilot boat was the right system to replace the *Peacock.* But implementing the first dedicated helicopter service on a pilotage ground in North America plus a radical change in pilot boat concept required the clearing of two major hurdles: the lack of consensus within the bar pilots' own ranks and securing the necessary funding in a rate proceeding that would be vigorously contested by the CRSOA, the organization representing the largely foreign flag ships calling on Columbia River ports. It took more than a year to secure the support of over half of the 24 bar pilots for the new transportation system. But industry support was another matter. Bar pilot president Captain Roger Nelson and vice president Captain Gary Lewin negotiated over several months with the CRSOA and the Port of Portland, but were unable to secure any commitment to the significant increase in pilotage rates necessary to fund the new system.

In the United States, pilot organizations are heavily regulated with rates established by a state agency just like those of a public utility. In Oregon, that agency is the Oregon Board of Maritime Pilots. In mid-1998, the bar pilots initiated what became an historic rate proceeding, both in terms of scope and result. No pilotage rate proceeding in Oregon history has consumed as many hearing days or accomplished such dramatic change in the character of a pilotage system. The bar pilots called 14 witnesses and presented over 1,500 pages of written testimony and exhibits over the course of eight days. The core issue was the proposed change in transportation system from a large station boat/daughter boat system to the combination of a helicopter and fast, maneuverable pilot boat.

In terms of cost, the dedicated helicopter was the big ticket item, adding an additional $1.5 million to an $8 million annual tariff that would be partially offset by a reduction in the number of bar pilots from 22 to 20, reflecting some of the additional efficiencies in the new system.

As the proponent of a change in the tariff funding their pilotage system, the bar pilots shouldered the burden of proof in a trial-like proceeding before Administrative Law Judge Ruth Crowley, whose job it was to analyze the evidence and prepare a comprehensive proposed order for the Oregon Board of Maritime Pilots.

In presenting their case, the bar pilots' evidence focused on two major themes: improving safety and increasing system efficiency.

From the standpoint of safety, the evidence in support of the helicopter was compelling. It was undisputed that the reason for the low levels of bar closures during the 1980s and 1990s was the willingness of the bar pilots to use the *Peacock's* daughter boat system to board and debark vessels in seas up to and exceeding 20 feet. But the bar pilots' devotion to providing pilotage service in such dangerous conditions imposed a heavy price in terms of injury and potential death. Nowhere else in the world did pilots board or disembark in such difficult conditions and the bar pilots' accident rate bore that out.

Captain Gerbrandt Van Santen, an active deep-sea pilot in the North Sea and manager of the organization providing Dutch maritime pilots to vessels transiting the English Channel and North Sea bound for ports in Europe, testified that helicopter embarkation of pilots in the Netherlands had been in place for 30 years and had grown to account for nearly one-third of the pilot transfers for vessels bound for Rotterdam. In those three decades, there had been only one accident and no injuries.

Captain Roger Nelson, a veteran of 20 years with the bar pilots and 18 years as a merchant mariner after graduating from the U.S. Merchant Marine Academy in Kings Point, New York, described the unique history of the Columbia River Bar Pilots, the extraordinary winter conditions in which they work, and the multi-year efforts to select the right pilot transfer system to replace the remarkable *Peacock* and her daughter boat. In his testimony, Captain Nelson made effective use of photos and riveting video of Captain Nelson himself holding a lanyard on the bow of the daughter boat and riding up a 20-foot swell toward a pilot ladder shifting in the strong winds. Making a split second decision that the daughter boat was near the top of the swell and that the pilot ladder was close enough, Nelson jumped to the ladder and scrambled up several rungs before looking back to verify that the daughter boat was away safely. For anyone seeing that videotape, there could be little doubt that the job of a bar pilot would qualify as one of America's "most dangerous."

While the white-haired Captain Nelson served as the primary historian in the 1998–99 rate proceeding, Captain Gary Lewin laid the foundation for a succession of witnesses showing how the helicopter/fast boat system would both dramatically improve pilot safety and enhance operational efficiencies not only for delivering pilots to and from vessels but to ships calling on the Columbia River. Captain Lewin presented the helicopter's impressive safety statistics on pilotage grounds throughout the world and the success both in terms of safety and operational efficiencies achieved during the four-month helicopter test on the Columbia River Bar in 1996–97.

Helicopter aviation experts with wide experience in hoist operations described the broad range of conditions in which the helicopter performed successfully during the four-month test and the safety features of the dual engine helicopters required for any operation involving the hoisting of personnel. According to these witnesses, the bar pilots should expect an accident-free operation from any reputable helicopter service provider. Captain Rod Leland, the former commanding officer of the U.S. Coast Guard Air Station at Astoria and

DELIVERING THE PILOT

The *Seahawk*, an Agusta twin engine helicopter, prepares to hoist Captain Robert Johnson down to an oil tanker. The bar pilot is clipped into a harness and winched down by wire cable to the deck of the ship. The winchman can be seen leaning out the door of the helicopter.

a helicopter pilot with over 20 years' experience, testified that he considered the combination of a helicopter and pilot boat to be the "safest of all possible systems."

For its part, the CRSOA opposed the extra cost of helicopter service, arguing that a test of four months was inconclusive and that the shift from a purely boat-based system should not be approved unless it could be accomplished at no extra cost. Curiously, the CRSOA had no evidence to confront that of the bar pilots that the cost of a station boat system with its greater number of bar pilots was actually more expensive than the switch to the helicopter/fast boat system. In anticipation of the potential approval of the new system, the bar pilots had not replaced two recent retirees, dropping their total number from 22 to 20 pilots. Even with the reduced number of 20 pilots, the speed of the new system was projected to increase the number of assignments per year per pilot by as much as 15 percent. Thus, a modest increase in vessel traffic would save another two pilot positions and cover $1 million of the projected $1.5 million in additional costs for the new system.

The Port of Portland participated actively in the rate proceeding, but ultimately took a neutral position in its closing brief, urging the Oregon Board of Maritime Pilots to "choose the system that best meets the pilots'

PICKING UP THE PILOT

Captain Bill Worth piloted the *Celebrity Millennium* from Astoria to sea and is being picked up by the *Seahawk* from the bow of the ship. *Photo by Michael Ciaglo.*

TODAY'S *COLUMBIA*

The newest *Columbia,* the fourth pilot boat bearing the name, is departing a log carrier after having transferred a bar pilot for an outbound assignment. *Photo by Robert Johnson.*

THE *CHINOOK*

The pilot boat *Chinook,* launched in 2000, approaches a container ship to disembark a bar pilot at sea. *Photo by Robert Johnson.*

MODERN-DAY PILOT TRANSFER

Captain George Waer is headed up the pilot ladder for an assignment after transferring from the *Chinook*. Note the cutaway slot between the black rubber bumpers where the pilot boat attempts to position the ladder during the transfer process. *Photo by Robert Johnson.*

needs," provided it was safe and preserved a reasonable cost structure.

After analyzing the voluminous record, Judge Crowley issue a 30-page proposed order recommending a 20 percent rate increase to fund the entire modernization of the pilotage system on the Columbia River Bar. On March 24, 1999, the Oregon Board of Maritime Pilots adopted Judge Crowley's proposed order with minor changes. In explaining its rationale to approve a complete transformation of the system and the first helicopter operation in the North America, the Board noted: "We defer to the bar pilots' professional judgment that the system can be safe and reliable. It is the pilots' lives at stake, after all, in the change in transportation system."

Evergreen Helicopters of McMinnville, Oregon, was the successful bidder for the initial three-year contract. Using an Italian Agusta K2 twin engine helicopter, the first transfer was made on August 6, 1999. The helicopter is crewed during the day with a pilot and winch operator, and at night a co-pilot is added. To maximize safety in the winching of personnel, FAA regulations require that the helicopter be capable of flight even with the loss of one engine.

In 2002, after the expiration of the original three-year contract, a new model Agusta helicopter, a 109E Power, began operations under a 10-year contract. Operated primarily by Arctic Air Service of Santa Maria, California, the Agusta helicopter has performed well in rigorous conditions, but exposure to sea spray and air has upped maintenance costs. In 2011, during a 25-day major inspection and maintenance period, corrosion in the helicopter tail boom was so extensive that much of it had to be replaced. Since operations first began in August 1999, the helicopter has transferred Columbia River Bar Pilots to and from ships in the Pacific Ocean over 25,000 times in 12 years. As this book goes to press, not a single injury accident has occurred.

Also approved in the 1998–99 rate proceeding was the second component of the pilotage system transformation: a fast jet-powered aluminum pilot boat designed by Alastair Cameron of Camarc Ltd. in the United Kingdom. Designing a new pilot boat for the Columbia River Bar was a unique challenge because of the requirement that the vessel have self-righting capability just like the *Peacock*. Built at Kvichak Marine Industries in Seattle, Washington, this new pilot boat was christened the *Chinook* and launched in August 2000. She was a dramatically different pilot boat than the now-retired steel-hulled *Peacock*, which was 87.9

feet in length, weighed 100 tons, and had a maximum speed of 16 knots.

At 72 feet and just 50 tons, the *Chinook* was nearly twice as fast at 30 knots and designed to serve a different function than a large station boat, instead working in tandem with the helicopter as part of a rapid pilot transfer system. The *Chinook* also added several safety features not found on the *Peacock* including a specialized handrail around the bow enabling a crewman to clip in and assist in boarding and debarking pilots and two man-overboard systems: a stern-mounted cradle that could scoop an unconscious person out of the water and a hydraulic davit on the starboard side for hoisting someone who was conscious.

Remarkably, the fifth grade class at Robert Gray Elementary School in Astoria suggested the name *Chinook* for the new pilot boat in a naming competition sponsored by the bar pilots. In relying on those fifth graders and christening their next generation pilot boat *"Chinook,"* the Columbia River Bar Pilots marked three symbolic links to the past: over 200 years since the 1792 discovery of the Columbia River by Captain Gray; over 180 years since Chinook Chief Concomly became the first officially-recognized pilot at the mouth of the Great River of the West; and 15,000 years since the battle of the great beaver Wishpoosh and the animal king Coyote produced the Missoula floods, the Chinooks and the other Columbia River Tribes.

With the launch of the *Chinook,* the bar pilots were on the cutting edge of pilot boat technology in the world, continuing a trend toward greater speed and maneuverability. One of the primary reasons for this trend was the movement of pilot boarding areas further offshore on many of the heavily trafficked pilotage grounds throughout the world. Moving that place further offshore where pilots were either transferred to an incoming vessel or taken off an outgoing vessel helped reduce congestion in constricted areas and provided adequate time for the master/pilot exchange and pre-passage planning before entering the generally confined and often congested channel on the pilotage grounds. Using the enhanced speed of the helicopter/fast boat system, the bar pilots moved their vessel marshaling area four to ten miles farther offshore from where it had existed for many decades before 1999.

This marshaling area offshore of the Columbia River entrance is now a legal requirement. In 2011, the Oregon Legislature expanded the seaward reach of the Columbia River Bar pilotage ground to five miles offshore and established a "precautionary zone" out to the 12 nautical mile U.S. territorial sea boundary line. No ship may enter or depart the Columbia River Bar precautionary zone without a pilot or in accordance with radio instructions from a bar pilot.

The Columbia River Bar Pilots were first exposed to the Camarc design in early 1998 when Captains Gary Lewin and Wayne Stolz traveled to the Netherlands to inspect and ride aboard a Camarc pilot boat, one of 10 delivered to the Dutch pilots in Rotterdam. Following the delivery of the *Chinook* in 2000, the trend toward fast pilot boats accelerated throughout the world. In the space of one decade, Camarc-designed pilot boats were built and delivered to eight other pilot organizations in North America.

But despite the major advances in pilot transfer technology reflected in the helicopter and *Chinook,* the bar pilot's job remains extremely dangerous. More than 80% of the transfers by helicopter are not land-on-deck operations, but complete the bar pilot's commute to or from an assignment in the Pacific Ocean at the end of a wire cable. And when the helicopter is unavailable due to fog or low visibility conditions, boarding or debarking via pilot boat in rough seas is treacherous.

On January 9, 2006, tragedy struck once again in a pilot ladder accident. Captain Kevin Murray was disembarking the outbound log carrier *Dry Beam* shortly before midnight. Just as he was descending that last 15 to 20 feet where the pilot literally rappels using one or two manropes, the *Chinook* bounced away from the side of the vessel. Captain Murray tried to stop his descent and step back to the ladder, but he lost his hold, and slid down the manropes and into the sea. He was wearing a float coat, but his personal marker light, one designed to activate a strobe upon emersion in seawater, was turned to the "off" position and did not deploy.

The *Chinook* crew searched frantically for Captain Murray, sighted him with a handheld spotlight, but then lost sight of him. When he reappeared five to ten minutes later, he appeared unconscious, floating face down. The *Chinook* attempted to retrieve him with the man overboard rescue basket, but the crew was unable to pull him into the basket and hold him in the rough seas.

An intensive search continued throughout the next day, but there was no sign of Captain Murray. Two days later, his body washed ashore at Roosevelt Beach, Washington, some 70 miles north of the mouth of the Columbia River.

The tragic loss of Captain Murray led to a comprehensive review of the bar pilots' safety procedures. Communications and man overboard drills procedures were enhanced. All pilots were required not only to have an automatically activating strobe light, but a personal locator beacon device called an EPIRB (Electronic Positioning Indicator Radio Beacon). Training procedures were beefed up and the *Chinook's* lighting was improved to enlarge the area of the sea that could be illuminated in searching for a man overboard.

When it came time to replace the last of the bar pilots' station boats, the 30 year-old third *Columbia,* the hard lessons of the tragic loss of Captain Murray resulted in multiple enhancements to the safety systems of the new fourth *Columbia,* which was christened in April 2008. These included an infrared camera for use in detecting a warm body out in the ocean plus two additional floodlights for man-overboard detection. The communication system for the rescue station at the stern of the vessel was upgraded, and there were multiple places where boat operator visibility was improved over the *Chinook* design. The rescue basket was also deepened by six inches to prevent it from moving over the top of a pilot in the water.

With the 2008 delivery of the fourth *Columbia,* the technological transformation of the pilotage transportation system serving the Columbia River Bar was complete. The *Chinook* and the *Columbia* work in tandem together, working alternate weeks. Each vessel also covers for the other during the two- to four-week period that each pilot boat spends in the shipyard for regular annual maintenance. By alternating their use week to week, the bar pilots predicted that the useful life of each vessel would extend to a full 15 years, five more years than that originally predicted by designer Cameron.

Entering the twenty-first century, the bar pilots' transportation system was truly one-of-a-kind in the world. Nowhere on the planet was a helicopter matched with a self-righting jet-powered pilot boat. But it was an appropriate pairing: the waterway spawning conditions posing the greatest risk of losing an oceangoing freighter was best tamed by a pilotage system capable of deploying skilled pilots safely in the broadest possible range of conditions.

EPILOGUE

THIS IS A BOOK THAT NEEDED TO BE WRITTEN. Time and again, history is lost or forgotten. Even more often, there is a kind of mindless complacency that lulls people into thinking that modern technology has produced a world where transportation systems are not only more efficient, but uniformly safe anytime, anywhere.

In the consideration of expensive safety systems in the transportation sector – whether motor vehicle, maritime, or aviation – there is always pressure to defer or avoid the extra cost of enhanced safety. I've seen this firsthand in a series of rate proceedings and other regulatory battles before the Oregon Board of Maritime Pilots over the last two decades. When it comes to the pilotage system on the Columbia River, the shipping industry has repeatedly – three times in the last decade alone – sought to consolidate the two pilotage systems serving vessels on the Columbia River: the Columbia River Bar Pilots at the mouth of the river and the Columbia River Pilots upstream from Astoria to multiple upriver ports. Without any supporting study, the argument is made that consolidation of these two pilot organizations will achieve multiple efficiencies, reduce costs, and eliminate the competitive advantage enjoyed by Puget Sound ports where the short runs to the deep water ports of Seattle and Tacoma require only one pilot.

In advancing this consolidation idea, which generally strikes a positive initial note – who can be against consolidation that reduces costs – the argument has been made that the status of the Columbia River Bar as one of the world's most dangerous waterway entrances is nothing more than "urban legend." The Columbia River Steamship Operators Association has made exactly this claim in one of its legal briefs. The scientific evidence assembled in this book definitively rebuts that charge: the Columbia River Bar is indeed the most dangerous entrance to a major commercial waterway in the world.

Why does this matter? In an era where the world has no tolerance for major transportation casualties, the critically important role of expert human skill sets must be recognized and respected. In the aftermath of the catastrophic *Exxon Valdez* oil spill in 1989, Congress rewrote antiquated oil spill liability statutes, and multiple states including Oregon upgraded their pilotage statutes to make the use of a local pilot compulsory rather than optional for oceangoing vessels. In its exhaustive 1994 study entitled *Minding the Helm,* the Marine Board of the U.S. National Research Council perhaps said it best:

> *Marine navigation and piloting occur in an operating environment characterized by extreme reliance on human performance, considerable diversity in geographic and hydrographic features, and great variability in operating conditions.*

The Columbia River amply demonstrates the extraordinary differences that can exist between pilotage grounds on a single river system. Piloting the Columbia River Bar is primarily a job that requires the skill of a seasoned ship master experienced in handling deep draft vessels in heavy seas. In contrast, the river pilot directing a vessel's transit within the Columbia River's narrow 600-foot navigation channel requires extensive knowledge of the river's currents, turns, shoals, and docks.

Captain Mitchell Stoller, a former Exxon oil tanker captain and L.A. Harbor pilot who also served a term on the National Advisory Safety Council, described the striking difference between the skill sets necessary for river and bar pilots this way: "Comparing a blue-water ship captain to a brown-water tug captain is similar to comparing a 747 pilot to a helicopter pilot. The pilots are not interchangeable between such different vessels and operating conditions."

The bottom line is that while extraordinary skill is required to avoid a tragic accident only in rare circumstances, regulators must continue to insist on licensure standards designed to ensure that those in the most critical transportation positions are as highly skilled as possible. And this mandates highly specialized expertise for both aviation and maritime pilots. Captain Chesley "Sully" Sullenberger's successful ditching of a commercial airliner in the Hudson River in January 2009 and the collective efforts of three bar pilots in turning the *Tai Shan Hai* incident in 2002 into a near miss rather than a major oil spill are two modern examples of the type of talent and expertise that are absolutely vital to reducing major transportation casualties to near zero.

The history of dual pilotage on the Columbia River – former ship captains at the bar and former tug captains on the river – dates back to before statehood and was first enacted into statute in 1860. The exemplary safety record that this dual pilotage system has delivered to the Pacific Northwest has been aided by improvements in technology, but its successful implementation is primarily dependent on "human performance." It is my hope that this book helps the public and policymakers alike to understand and appreciate the unique and critical role of the Columbia River Bar Pilots to the navigational history of the Columbia River Bar and to its future safety.

ROSTER OF BAR PILOTS

LAST NAME	FIRST NAME	ORGANIZATION	YEAR BEGAN	YEAR RETIRED	YEARS SERVED
Alexander	Robert	Self	1847		
Anderson	Charles	Bar Pilots	1917	1948	31
Anderson	Gustaf	Port of Portland	1914		
Ash	Clarence	Bar Pilots	1926	1955	29
Astrup	Hagen	Port of Portland	1915		
Aune	Harold	Bar Pilots	1948	1961	13
Aursland	Kell	Bar Pilots	1983	2003	20
Bailey	Lester	Puget Sound Tugboat	1886		
Barrett	Barry	Bar Pilots	1990		
Belmont	Andrew	George Flavel	1853		
Bochau	William	Portland Tugboat Co.	1883		
Brown	Hugh	Bar Pilots	1952	1982	30
Bruneau	Joseph	Bar Pilots	1967	1983	16
Campbell	James E.	Asa Simpson	1881		
Cann	Archie	Port of Portland	1909		
Carruthers	Robert	Washington Pilot Board	1896		
Cathcart	Virgil	Bar Pilots	1963	1987	24
Collamore	Ellwood	Bar Pilots	1983	2008	25
Collins	Raymond	Bar Pilots	1960	1983	23
Concomly	Chief	Self	1813	1830	17
Conway	George J.	Bar Pilots	1917	1960	43
Cordiner	Peter	State of Oregon	1889		
Corno	Paul	Self	1865		
Crocker	James S.	George Flavel	1878		
Crosby	Alfred	George Flavel	1853		
Dempsey	Deborah	Bar Pilots	1994		
DeSassise	John F.	Bar Pilots	1957	1967	10
Dickinson	Dale	Bar Pilots	1962	1986	24
Dillon	Michael	Bar Pilots	1989	2004	15
Doig	Thomas D.	George Flavel	1872		
Dooney	Kevin	Bar Pilots	2008	2011	3
Dougan	William	Bar Pilots	1959	1976	17
Duerr	Larry	Bar Pilots	2008		
Edwards	Charles	Jackson Hustler	1849		
Elsensohn	Robert	Bar Pilots	1968	1982	14
Erbe	George	Bar Pilots	1939	1967	28
Farnsworth	Asa Cole	George Flavel	1853		
Farrer	Erwin	Asa Simpson	1879		
Faulkner	Warren W.	Bar Pilots	1961	1983	22
Ferchen	Peter	George Flavel	1855		
Flavel	George	George Flavel	1850	1887	37

LAST NAME	FIRST NAME	ORGANIZATION	YEAR BEGAN	YEAR RETIRED	YEARS SERVED
Flavel	George C.	George Flavel	1877		
Gage	John W.	Asa Simpson	1869		
Geer	George	Self	1847		
Gibson	Robert W.	Bar Pilots	1964	1990	26
Gillard	Frank	Bar Pilots	1948	1967	19
Gillette	Elton	Bar Pilots	1941	1957	16
Gillman	M.M.	Asa Simpson	1870		
Gladwell	Thomas	Asa Simpson	1859		
Glick	Michael	Bar Pilots	1999		
Graham	Dan	Union Pacific	1889		
Grant	Thomas	Portland Tugboat Co.	1877		
Grassman	Johan	State of Oregon	1890		
Gray	John Henry Dix	Ilwaco Steam Navigation	1876		
Gresham	William	State of Oregon	1890		
Grunstad	Einar	Union Pacific	1917		
Gunderson	Charles	Asa Simpson	1881		
Hall	William	Bar Pilots	1925	1945	20
Hammon	Alf	Bar Pilots	1961	1986	25
Hampson	George	Bar Pilots	1924	1934	10
Hansen	Charles E.	Asa Simpson	1883		
Hansen	Hans	Bar Pilots	1940	1951	11
Hanson	Arthur	Bar Pilots	1923	1951	28
Harriman	Cyrus	Port of Portland	1912		
Harriman	Joseph	Union Pacific	1890		
Hatfield	Job	Jackson Hustler	1849		
Hawks	Thomas	Self	1850		
Hewett	A. W.	Asa Simpson	1878		
Hill	James	Asa Simpson	1862		
Hilton	Eli	Asa Simpson	1875		
Hirsch	Frederick	Bar Pilots	1901	1937	36
Hobson	Richard	Self	1876		
Howes	Richard E.	State of Oregon	1879		
Hubbard	L. H.	Union Pacific	1873		
Hustler	Jackson G.	Jackson Hustler	1849		
Jackson	Paul	Bar Pilots	1986	2003	17
Jerrell	Fred	Bar Pilots	1973	1993	20
Jessen	Gus	Asa Simpson	1879		
Johnson	Charles F.	State of Oregon	1873		
Johnson	Charles O.	Union Pacific	1915		
Johnson	Erick	Asa Simpson	1875		
Johnson	H.A.K.	Asa Simpson	1867		
Johnson	James	Self	1848		
Johnson	Phillip	Asa Simpson	1891		
Johnson	Robert	Bar Pilots	1986		
Johnson	Thomas	State of Oregon	1888		
Johnston	Ronald	Bar Pilots	1982	2004	22
Jordan	Daniel	Bar Pilots	2004		
Kofoed	Charles	George Flavel	1865		
Lane	Charles	Bar Pilots	1984	2008	23
Langkilde	Alvin	Bar Pilots	1925	1946	21
Langkilde	August	Bar Pilots	1942	1948	6

LAST NAME	FIRST NAME	ORGANIZATION	YEAR BEGAN	YEAR RETIRED	YEARS SERVED
Lapping	John	Union Pacific	1913		
Latham	Thomas	Asa Simpson	1879		
Lattie	Alexander	Hudson's Bay Co.	1844		
Leighton	Arthur	Union Pacific	1901		
Lessard	James	Bar Pilots	1979	1999	20
Lewin	Gary	Bar Pilots	1983		
Lofstedt	August	Bar Pilots	1920	1944	24
Magee	James	Asa Simpson	1887		
Malcolm	Alexander	State of Oregon	1879		
Marquart	Robert	Bar Pilots	1935	1967	32
Mathews	Henry	Asa Simpson	1880		
Matteo	Phillip	Bar Pilots	2008		
McAlpin	Kenneth	Bar Pilots	1957	1983	26
McAvoy	James	Bar Pilots	1975	1994	19
McVicar	Dan	George Flavel	1869		
Metzger	William	George Flavel	1860		
Mouatt	William	Hudson's Bay Co.	1848		
Murray	Kevin	Bar Pilots	2004	2006	1
Nehring	Curtis	Bar Pilots	2005		
Neill	Thomas	State of Oregon	1893		
Nelson	Roger	Bar Pilots	1979	2005	25
Nemecek	Eugene	Bar Pilots	1976	1997	21
Nielson	Anton	Puget Sound Tugboat	1915		
Nolan	Michael	Bar Pilots	1903	1939	36
Olsen	Charles	State of Oregon	1888		
Olsen	Chris	Asa Simpson	1899		
Olsen	Henry	State of Oregon	1883		
Olsen	William	State of Oregon	1888		
Parker	Karl Page	Bar Pilots	1943	1964	21
Parsons	Ed	Puget Sound Tugboat	1918		
Pease	George	Portland Tugboat Co.	1894		
Peterson	Oluf	State of Oregon	1886		
Pickernell	John	Self	1842		
Plumlee	James W.	Bar Pilots	1973	2002	29
Powers	Thomas	State of Oregon	1888		
Quinn	Edgar	Bar Pilots	1951	1973	22
Randall	Samuel	Asa Simpson	1886		
Rankin	Oliver	Bar Pilots	1923	1939	16
Reed	John	Asa Simpson	1874		
Reed	Charles	Bar Pilots	1950	1976	26
Richards	James	Bar Pilots	1981	2007	26
Richardson	Charles	State of Oregon	1883		
Richter	Joseph	State of Oregon	1888		
Ridenour	Walter	Bar Pilots	1967	1981	14
Riggs	Thron	Bar Pilots	1993		
Rogers	Moses	Paul Corno	1865		
Salfen	Jeffrey	Bar Pilots	1977	2006	29
Sanders	Charles	Asa Simpson	1885		
Sayer	Stanley	Bar Pilots	1975	1994	19
Scarborough	James	Hudson's Bay Co.	1844		
Schroeder	Harvey	Bar Pilots	1945	1976	31

LAST NAME	FIRST NAME	ORGANIZATION	YEAR BEGAN	YEAR RETIRED	YEARS SERVED
Sherman	Fred	George Flavel	1881		
Snugg	Gus	Puget Sound Tugboat	1881		
Staples	Leo	Asa Simpson	1891		
Staples	Marshall D.	George Flavel	1876		
Stolz	Wayne	Bar Pilots	1994		
Stone	Geoffrey	Bar Pilots	1980	2007	27
Stroup	James	Bar Pilots	1967	1998	31
Swanson	Reinert	Bar Pilots	1909		
Tatton	James	State of Oregon	1902		
Thomsem	Daniel	Puget Sound Tugboat	1910		
Tierney	Michael	Bar Pilots	2006		
Torjusen	John	Bar Pilots	2002		
Tribble	William	Asa Simpson	1878		
Waer	George	Bar Pilots	1981		
Wass	Andrew	George Flavel	1869		
West	Martin	Bar Pilots	1966	1991	25
White	Cornelius	Jackson Hustler	1849		
Williamson	Richard	Bar Pilots	1955	1963	8
Wood	George Warren	Bar Pilots	1883		
Wood	Kenneth P.T.	Bar Pilots	1920		
Worth	William	Bar Pilots	1992	2011	19

INDEX